CAMBRIDGE LIBRARY COLLECTION

Books of enduring scholarly value

Technology

The focus of this series is engineering, broadly construed. It covers technological innovation from a range of periods and cultures, but centres on the technological achievements of the industrial era in the West, particularly in the nineteenth century, as understood by their contemporaries. Infrastructure is one major focus, covering the building of railways and canals, bridges and tunnels, land drainage, the laying of submarine cables, and the construction of docks and lighthouses. Other key topics include developments in industrial and manufacturing fields such as mining technology, the production of iron and steel, the use of steam power, and chemical processes such as photography and textile dyes.

The Design and Construction of Harbours

Thomas Stevenson (1818–1887) was the son of the engineer Robert Stevenson, and father of the writer Robert Louis Stevenson. Like his brothers David and Alan, he became a lighthouse designer, being responsible for over thirty examples around Scotland. Throughout his career he was interested in the theory as well as the practice of his profession, and published over sixty articles on engineering and meteorology. He was an international expert on lighthouses and harbour engineering. This work was first published in 1864 as a development of his article on harbours in the eighth edition (1857) of the *Encyclopaedia Britannica*, and considerably expanded in a second edition of 1874 which is reprinted here. Stevenson studied how the wind, waves and tides would act on the coastline and man-made structures, and the design of each harbour needed to take a wide range of factors into consideration.

Cambridge University Press has long been a pioneer in the reissuing of out-of-print titles from its own backlist, producing digital reprints of books that are still sought after by scholars and students but could not be reprinted economically using traditional technology. The Cambridge Library Collection extends this activity to a wider range of books which are still of importance to researchers and professionals, either for the source material they contain, or as landmarks in the history of their academic discipline.

Drawing from the world-renowned collections in the Cambridge University Library, and guided by the advice of experts in each subject area, Cambridge University Press is using state-of-the-art scanning machines in its own Printing House to capture the content of each book selected for inclusion. The files are processed to give a consistently clear, crisp image, and the books finished to the high quality standard for which the Press is recognised around the world. The latest print-on-demand technology ensures that the books will remain available indefinitely, and that orders for single or multiple copies can quickly be supplied.

The Cambridge Library Collection will bring back to life books of enduring scholarly value (including out-of-copyright works originally issued by other publishers) across a wide range of disciplines in the humanities and social sciences and in science and technology.

The Design and Construction of Harbours

A Treatise on Maritime Engineering

Thomas Stevenson

CAMBRIDGE UNIVERSITY PRESS

Cambridge, New York, Melbourne, Madrid, Cape Town,
Singapore, São Paolo, Delhi, Tokyo, Mexico City

Published in the United States of America by Cambridge University Press, New York

www.cambridge.org
Information on this title: www.cambridge.org/9781108029674

This edition first published 1874
This digitally printed version 2011

ISBN 978-1-108-02967-4 Paperback

THE
DESIGN AND CONSTRUCTION
OF HARBOURS

THE

DESIGN AND CONSTRUCTION

OF

HARBOURS

A TREATISE ON MARITIME ENGINEERING

BY

THOMAS STEVENSON, F.R.S.E., F.G.S.

MEMBER OF THE INSTITUTION OF CIVIL ENGINEERS, AND
AUTHOR OF 'LIGHTHOUSE ILLUMINATION,' ETC.

SECOND EDITION

EDINBURGH

ADAM AND CHARLES BLACK

MDCCCLXXIV

Printed by R. & R. CLARK, *Edinburgh.*

PREFACE TO SECOND EDITION.

HAVING been informed by the publishers that a new edition of this work was frequently asked for, I have revised it for republication with as much care as possible.

The reader will observe that many additional subjects have been introduced, while most of the chapters have been very considerably extended; and I have again to thank my professional brethren for the assistance which they have rendered.

EDINBURGH, *February* 1874,
84 GEORGE STREET.

PREFACE TO FIRST EDITION.

THE following pages are a reprint, with many additions, of the Article "Harbours," in the last edition of the Encyclopædia Britannica. The late Professor Hosking, of the London University, proposed to republish that Article, along with several others, from the Encyclopædia, as an independent Treatise on Civil Engineering and the Constructive Sciences, and it was accordingly revised, with the intention of its appearing as one of the proposed series. Owing to the death of Professor Hosking, that work was not proceeded with, and the Article is now published by itself in a separate form.

I have endeavoured to make this volume a useful contribution to Maritime Engineering, by introducing into it, as largely as possible, the results of actual practice. In requesting a measure of indulgence for the insertion of matter which is familiar to the experienced practitioner, I would remind the reader of the necessarily comprehensive character of an

Encyclopædia article; while, for imperfections and undue abbreviations, I can only urge my desire to enlarge upon those branches of which I happened to have had most personal experience, rather than to deal too much with subjects at second hand. The reader will, however, notice how much I am indebted to several of my engineering friends, and also to the excellent "Cours de Construction—Ouvrages Hydrauliques des Ports de Mer," by M. Minard.

EDINBURGH, *March* 1864.

CONTENTS.

CHAPTER I

INTERIOR WORKS—EXTERIOR WORKS—DIFFERENT CLASSES OF HARBOURS.

CHAPTER II.

GEOLOGICAL AND OTHER PHYSICAL FEATURES.

CHAPTER III.

GENERATION OF WAVES.

CHAPTER VI.

DESIGN OF PROFILE, ETC., OF HARBOURS IN DEEP WATER.

CHAPTER VII.

DESIGN OF PROFILE, ETC., FOR TIDAL HARBOURS.

CHAPTER VIII.

DESIGN OF GROUND-PLAN OF HARBOURS.

CHAPTER IX.

DOCKS, TIDE-BASINS, LOCKS, GRAVING-DOCKS, SLIPS, ETC.

CHAPTER X.

MATERIALS, KINDS OF MASONRY, IMPLEMENTS, ETC.

CHAPTER XI.

ON THE EFFICACY OF TIDE AND FRESH WATER IN PRESERVING THE OUTFALL OF HARBOURS AND RIVERS.

CHAPTER XII.

MISCELLANEOUS SUBJECTS RELATING TO HARBOURS.

CHAPTER I.

INTERIOR WORKS—EXTERIOR WORKS—DIFFERENT CLASSES OF HARBOURS.

1. Harbours of Refuge and Anchorage Breakwaters—2. Deep-water and Tidal Harbours for commercial purposes—3. Kanted or Curved Piers—Free End of Piers and Breakwaters—4. Straight Piers—5. Quay or wharf—Different Exposures require different kinds of Harbours—Requirements of a good Harbour—Technical Terms.

ALL harbours may be classed either as havens for the protection of ships during storms, or as ports adapted for commercial purposes.

Of the first-mentioned class, or those which are called harbours of refuge, some are natural and some artificial. Many parts of the British coasts are amply provided with natural bays and creeks, while in other districts the accommodation and shelter have been entirely supplied by artificial means. Thus, great portions of Ireland and of the west coast of Scotland are plentifully provided with excellent deep-water bays and anchorages; while on the east and south-west shores of Britain there are few natural harbours. Cromarty Bay, described by Buchanan as "adversus omnes tempestates portus salutiferus ac certum perfugium,"* is 200 miles distant from the Firth of Forth, which is the nearest natural harbour to the south; and there are no less than 400 miles between the Firth of Forth and the Thames, which may be considered as the next really unexceptionable place of refuge. On the west

* Rer. Scot. Hist., Auct. G. Buchanano, 1583.

coast, there are about 200 miles between the nearest harbours of Milford and Loch Ryan. The construction of artificial places of refuge becomes, therefore, a very important matter in a country where every winter's list of shipwrecks and loss of life reminds us how much nature has left for art to accomplish. For the most perfect body of evidence regarding the ports of Britain, we cannot do better than refer to the volumes of Reports by the Royal Tidal Harbours' Commission of 1846; for the completeness of which the public is mainly indebted to the zeal of the late Admiral Washington, the indefatigable hydrographer to the Admiralty.

The designing of harbours constitutes confessedly one of the most difficult branches of civil engineering. In making such a design, the engineer, of course, avails himself of the information which is to be derived from past experience, and endeavours, to the best of his power, to institute a comparison between the given locality and some existing harbour which he supposes to be similarly situated. Perfect identity, however, in the physical peculiarities of different localities, seldom, if ever, exists, and all that can be done in deriving benefit from past experience is to select the harbour which seems most nearly to resemble the proposed work.

In order the better to understand the nature of the difficulties which beset the marine engineer, let us suppose that he is called upon to design works for the accommodation of shipping in a given locality. The questions which immediately press on his attention are, *first*, What is the cheapest kind of design which is suitable for the place and sufficient for the class of shipping which has to be accommodated? and *second*, What are the smallest sizes of materials that are admissible in its construction? as on this the cost of the work may materially depend. Before considering how far it is possible to answer these questions, let us endeavour to define the different varieties of design into which all sorts of harbours may, with propriety, be resolved.

In the first place, they may be all classified under two main heads—viz. *Interior Works* and *Exterior Works.*

Interior Works.—The interior works are provided for the accommodation and repair of vessels. They consist of tide basins with or without gates—of wet docks with entrance locks—of graving (or dry) docks—patent slips—gridirons, etc., the three last mentioned being only for the reception of vessels requiring repairs. It is not proposed to enter on any very detailed description of the inner works of harbours, as, from their situation, they are necessarily protected from heavy seas, and are consequently more nearly of the same character at different harbours than the exterior works, the nature of which must vary more or less with every locality.

Exterior Works.—The exterior works of harbours may be conveniently enough classified under the following five different designs—

Different Classes of Harbours.

1*st, Harbours of Refuge and Anchorage Breakwaters,* consisting of one or more breakwaters, so arranged as to form a safe roadstead, which shall be easily accessible to the largest vessels in all states of the weather and tide. A breakwater forms a barrier either complete or partial to the progress of the waves, and is intended for sheltering the anchorage ground under its lee. It is not, like piers or quays, used for commercial traffic, and therefore a parapet is not necessarily required for preventing the waves from breaking over the top, although it may be useful as a protection against the wind.

2d, Deep-water and Tidal Harbours for commercial purposes (Fig. 1, *a* and *b*).—A harbour for commercial purposes is any arrangement of piers or breakwaters, or of both, which incloses and so tranquillises a sheet of water, that vessels may be moored at the quay walls or wharves, which form the inner sides of the piers. Where the coast-line lies open to a heavy sea it is often found necessary to make a double harbour (Fig. 1, *b*). In such a case the entrance to the inner basin is situated within the sheltered area formed by the outer works.

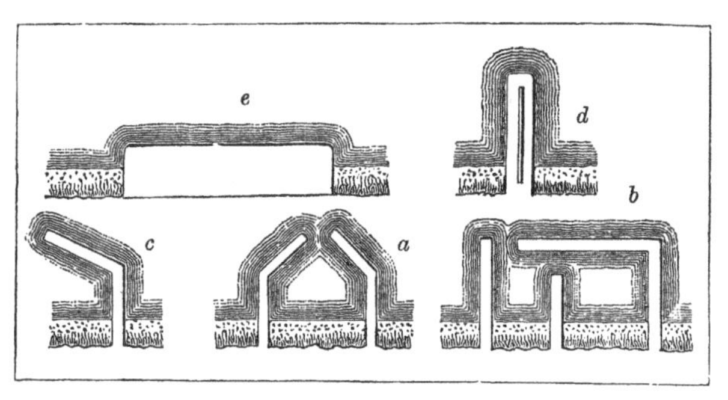

Fig. 1.

3d, Kanted or Curved Piers (Fig. 1, *c*).—Where there is a single pier of this kind, vessels lie under the lee of the kant or kants, and the sheltered side of the pier is therefore finished as a quay. The pier may, as for example that at Bournemouth, have a double kant, or cross-head, built at right angles to the main portion, so as to give the structure the form of the letter T; at one side or the other of which, according to the direction of the wind, vessels can always find some shelter. We propose to term the outer extremity of any single pier its *free end*, because there is an expanse of open sea all round it, in contradistinction to the outer end of a close harbour, where the sea-room is limited to the breadth of entrance, which is always kept as narrow as is consistent with the safe passage of vessels. Both extremities of all single isolated breakwaters are, of course, free ends, as are also the seaward ends of all single breakwaters which are connected with the land.

4th, Straight Piers (Fig. 1, *d*).—A straight pier generally projects at right angles to the coast line, with a free end at its seaward extremity, and, unless when the wind blows right in upon the shore, a straight pier will always afford some shelter on its lee side. In order to get the full advantage of this kind of pier, both sides are sometimes finished as quay walls, and the parapet, if there be one, is built in the middle of the roadway, as at Granton in the Firth of Forth.

5th, Quay or Wharf (Fig. 1, *e*).—A quay wall is usually built parallel to the line of shore. It affords no shelter of any kind, and the only advantage which it possesses is that of enabling vessels to load and unload without their having to "beach," or, where the shores are steep, even to take the ground. The same object may also be effected by an open framework of timber piles—by a suspension bridge, with a wharf at its outer end—or by a floating pier rising and falling with the tide, and connected with the shore by a bridge.

Different Exposures require different kinds of Harbours.—It will be observed that all the kinds of piers or harbours just enumerated differ materially from each other in the amount of shelter which they afford, and are therefore suitable for places having very different degrees of exposure. Now we find an almost infinite diversity in the height of waves when we pass along the coast from one part to another. In some places there are shores which lie open to the full fury of the ocean, while other parts of the same coast are protected in some directions by projecting headlands or islands. Then, leaving the main coast, we have the shores and bays of narrow sounds, the breadths of which vary at different places; and lastly, we have creeks so perfectly land-locked, as to afford complete natural shelter in the worst weather. In some situations the foreshore is steep, affording sufficient depth for heavy waves not only to reach the beach, but to tear up rocks at levels far above the high-water line; while in others it is so flat and shallow as to form a natural breakwater for the protection of the coast. In some districts there are tides rising forty or fifty feet, in others not as many inches; and lastly, we have differences in the geological formation and in the tendency to deposit. Now, it is quite as bad engineering to adopt the cowardly and unjustifiable policy of erecting in sheltered seas works that are heavy enough for the open ocean; as through ignorance or foolhardiness to fall into the opposite error of designing structures that are deficient in strength and efficiency. The very first step to be taken, therefore, is to select from the different classes of designs which have been enumerated, the one which is best adapted to the physical peculiarity of the situation. The engineer, in order to make this selection judiciously—keeping ever in view the essential elements of *stability*, *expense*, and *safe accommodation for the trade of the port*—must consider the following queries:—

First, Is the place so well sheltered naturally as to require no artificial protection of any kind, so that a quay without a parapet, or an open framework of timber, will be sufficient for vessels to lie alongside without risk of damage in all ordinary states of the weather? Examples of such quays may be found in rivers and creeks even where there is a considerable expanse of water, such as Greenock, Invergordon, and the like.

Second, Is the place situated in a Sound or Estuary, where the cross waves or those which come *endways* on the pier, are small, owing to the estuary being narrow, and where the heaviest waves are those which assail the work on its sides, so that a straight pier will be sufficient? of this Burntisland in the Firth of Forth is an example.

Third, Is it necessary to protect the berthage by means of a curved or kanted pier, as may be seen in a great variety of places where the sea is not very heavy?

Fourth, Is it necessary that a space of water should be inclosed between two piers inclined to each other till they nearly meet, and admitting (through the narrow entrance thus formed) only a small portion of the outside wave, which is afterwards reduced by expansion into the inclosed area? Examples of this may be seen at Ramsgate, and many other places on the coasts of Britain.

Fifth, Must we have recourse to what may be called a *compound harbour,* consisting of one harbour within another, where the outside waves are first reduced by expansion into the area of the outer or stilling harbour, after which a yet greater reduction is attained by a second expansion of a portion of the reduced wave into the area of the inner basin? Examples of such double harbours are common on all coasts which are much exposed.

The Requirements of a good Harbour.—After the engineer has satisfied himself as to the *general* character or class of design required,—which is undoubtedly the principal question to be settled, he must next consider the details. If the place be much exposed he must arrange the different parts of the work so as to produce a harbour which may be easily taken and left in stormy weather, without endangering the tranquillity of the internal area; for it is *the combination of the qualities of easy entrance and exit with a good "loose" and a smooth interior, which alone constitutes a good harbour.* Lastly, he must fix the width of the piers and height of the parapets, and assign the sizes and determine the arrangement of the constituent materials in such proportions as to ensure the stability of the whole structure.

What follows is an attempt to assist the engineer in the solution of some, at least, of these and other questions affecting the construction of harbours.

The local characteristics which at the outset demand our consideration are—1st, The geological and other physical peculiarities of the shore ; 2d, the exposure ; 3d, the force of the waves due to the exposure ; 4th, the strength, direction, and range of the tides ; 5th, the depth of water of the bay or sea in which the harbour is to be placed ; 6th, the proximity of deep water to the pier itself, which, of course, depends on the slope of the foreshore ; and 7th, the angle which the coast-line makes with the direction in which the heaviest waves come.

Before proceeding to consider these different subjects it may simplify matters to non-professional readers, or others who are beginning to study the subject, to denote, as shown in Figs. 2 and 3 on the opposite page, a few of the technical names of the different parts of a pier, which will be very frequently employed as we proceed.

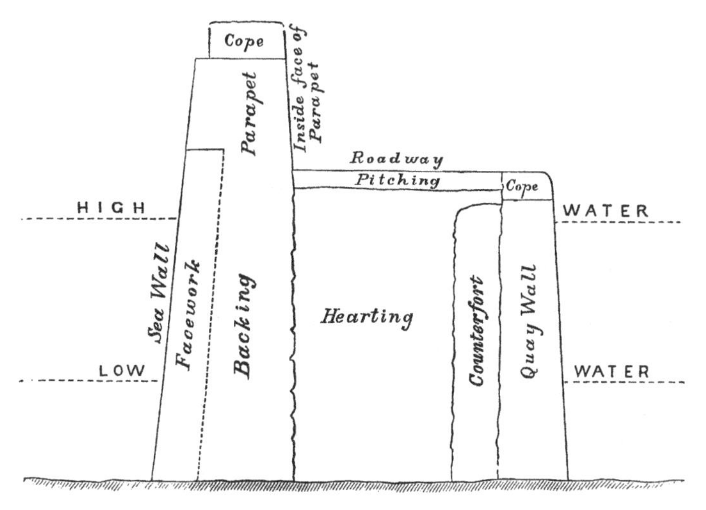
Cope
Parapet
Inside face of Parapet
Roadway
Pitching
Cope
HIGH
WATER
Sea Wall
Facework
Backing
Hearting
Counterfort
Quay Wall
LOW
WATER

Fig. 2.

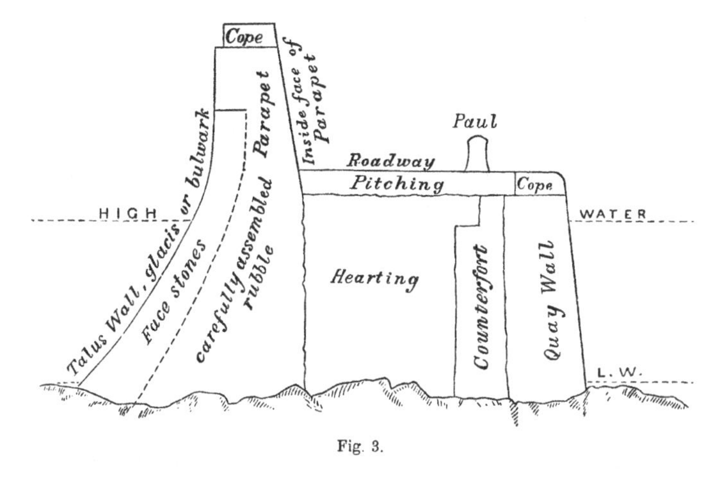
Cope
Inside face of Parapet
Parapet
Paul
Roadway
Pitching
Cope
HIGH
WATER
Talus Wall, glacis or bulwark
Face stones
carefully assembled rubble
Hearting
Counterfort
Quay Wall
L. W.

Fig. 3.

CHAPTER II.

GEOLOGICAL AND OTHER PHYSICAL FEATURES.

Apparent Signs of the existence of a heavy Sea due really to gradual Attrition —Appearances leading to an Under-estimate of the Exposure—Long Ledges of Rock sometimes dangerous — Levels of Vegetation and of Surf-marks deceptive—Great Storms as contrasted with heavy Gales—Tempest of 1703—Level assumed by Mud a measure of Exposure—Action of Waves at bottom—Level of Mud on German Ocean—Exposure of Dutch Coast.

THE engineer who is accustomed to contemplate the features of different sea-beaches is able to draw conclusions as to the exposure, the means of arriving at which cannot be imparted to another, but must be gained by personal experience alone. We shall therefore confine our remarks to one or two characteristics which admit of description. In some places these are obvious and easily judged of, but in others there are deceptive appearances which the engineer should be able to detect.

Apparent Signs of Effects of heavy Waves due to gradual Attrition.—Sir Henry de la Bêche mentions in his Geological Manual* that portions of the undercliffs at Pinhay, near Lyme Regis, have been destroyed, not so much by the action of the sea as by that of land-springs. The upper strata at that part of the coast are described as consisting of gravel, chalk, and greensand, resting on a bed of clay, resulting from the disintegration of the greensand beds above, and of the upper part of the lias beds beneath. The strata which lie

* Manual of Geology. London, 1832.

above the clay, being porous, admit of the percolation of the surface water, which descends till it reaches the impervious clay-bed, when it escapes by the easiest road to the sea. In escaping, it gradually washes away the outer parts of the clay-bed, thus depriving the upper strata of chalk and green-sand of their support. The under bed of lias, which resists the action of the land-springs, forms a base which receives the ruins of the upper strata as they fall, until the sea in front, having gradually removed the bed of lias that supports them, allows them to drop upon the beach, where they are broken up by the sea. The waves therefore perform only a very minor part in the destruction of the cliffs; the principal cause being the gradual escape of water from the land-springs above.

At some parts of the coast of Scotland there are similar apparent proofs of the waves having exerted a great force ; but the phenomena presenting these appearances, though no doubt indirectly due to the action of the sea, have not resulted from the stroke of any one wave, and form, therefore, no measure of its force. The coal formation which crops out along these shores consists of alternating beds of freestone and soft shale, in the latter of which, owing to its friable nature and open texture, the surf makes easy inroads by gradual attrition. The beds of sandstone, being of much harder texture, wear more slowly, and are therefore often seen projecting several feet beyond the subjacent shale. When the freestone beds have been sufficiently undermined, they break off *by their own weight,* and thus large tabular masses are detached, which, to a casual observer, present all the appearance of having been broken off by the impact of the waves, whereas the whole effect is due to the geological accident of the alternation of strata possessing different degrees of hardness. The disproportionate liability to corroding action produced by breakers when acting on rocks which are not of homogeneous texture and which vary in hardness, are not, however, confined to the sedimentary strata. Among the igneous rocks, Dr. Macculloch, in his Western Islands of Scotland, adduces the celebrated Fingal's Cave in Staffa as an example of this inequality of degradation. Dr. Macculloch considers that the formation of the cave, which is 130 feet long, is due to the numerous joints in the basaltic columns which confront the sea at that spot, while the general character of the adjacent trap is a rock presenting fewer seams to the action of the waves.*

Appearances leading to an Under-estimate of the Exposure. —There are, however, in other districts deceptive appearances of a different kind, which may betray the observer into the opposite and more dangerous error of under-estimating the exposure, or of misleading him as to the level reached by very high tides.

The engineer must remember, in drawing inferences from the rate of degradation, that this is not only dependent on the relative hardness of the rocks which are exposed to the waves, but also on the dip of the strata in relation to the direction of the breakers. Sir Henry de la Bêche says—"In many situations on the southern coasts of Devon and Cornwall, the slaty rocks dip in such a manner towards the sea that the waves have never effected more than the removal of some loose superficial matter, the same that covers all the hills in the vicinity. In fact a skilful engineer could not have protected the coast better than has been accomplished by the dip of the strata."†

Ledges of Rock dangerous.—This remark of Sir Henry's sug-

* Description of the Western Islands of Scotland, by John Macculloch, M.D., vol. ii. p. 11 : Lond. 1819.

† Manual of Geology, p. 71.

gests another source of danger which ought not to be overlooked. If a sloping direction of the strata has the effect of reducing the force against the coast by altering the direction of the surf, it is equally clear that where long ledges of rock cross the line of direction of a proposed pier, there may be expected an intensified action at the points of junction of the rock with the masonry. Long ledges of rock, though affording useful shelter where the works run parallel to them, are therefore sources of danger where this parallelism cannot be preserved in laying out the lines of the piers. All attempts to carry works across those long narrow chasms which separate rocky shelves, or even to cross creeks of considerable width, must be regarded as peculiarly hazardous, and special provisions are required for resisting the concentrated action which is common to these and all other places where the sea is *gorged.* It is therefore a fallacy to suppose that a chain of isolated outlying rocks necessarily furnishes facilities for the erection of a breakwater, and therefore tends to reduce the cost of erection. On the contrary, although the existence of such rocks may reduce the amount of diving-bell work, it may increase the risk of failure to a great extent.

It must, however, be understood that cases occur where, in order to find sufficient harbour room, or to effect some particular object, it becomes necessary to erect sea-works in situations where the pent up waves must be fully encountered. The sea-wall of the Victoria Harbour at Dunbar is an instance of this kind, for the basin that had to be enclosed forms the landward portion of a narrow creek. The outer wall has therefore not only to check, without any lateral relief, the *whole* of the waves, which formerly dashed into the creek, but owing to the outline of the coast it has also to encounter them nearly at right angles to their direction.

Levels of Vegetation and of Surf-marks deceptive.—Mistakes as to the level of the highest tides are sometimes made by drawing too hasty conclusions from the presence of vegetable life. I have seen the thrift or "sea pink" (*Armeria maritima*), which seems to indicate unmistakably the limit of the rise of the highest tide, submerged, even in calm weather, during equinoctial springs. The tide of 8th January 1868 rose at Leith, according to Mr. George Robertson's observation,* four feet four inches higher than the calculated height for the day, and higher than any that has occurred for the last eighteen years; and at Hull it rose five feet five inches above the calculated height.† Hence in defining any measurements which have the tide-level for a datum—as, for example, in clauses connected with river conservancy—it should always be stated that it refers to a given ordinary spring uninfluenced by wind. Where there is no opportunity of making tidal observations, the level of the *lepas* or barnacle shell, which is generally very sharp and well defined, may be adopted as a help in fixing the mean level. The greatest height at which I have noticed this shell-fish, varies from about half-way between the high-waters during neaps and ordinary springs, to about high-water of the highest neap tides or of the lowest springs. Nor must the existence of grass and other land-vegetation be regarded as any decisive proof that the surf never reaches it. In the Shetland Islands there may often be seen large blocks of rock (and to these we shall afterwards refer), which, during storms, have been driven over the land

* Proceedings Royal Society of Edinburgh, 1868, vol. vi. p. 296.

† Franklin mentions that a pond nine miles wide, and of an average depth of three feet, was acted on by a strong wind, which forced the water from one side, so that it was laid bare, and the depth of water on the other side was increased to six feet.—Rep. of the Commiss. of Agriculture for the year 1870, p. 610: Washington, 1871.

at heights much greater than the level at which vegetation commences, and far above the ordinary run of the surf.

Great Storms as contrasted with Heavy Gales.—I would also add a caution applicable specially to all inquiries regarding the occurrence of storms. It is a common and dangerous mistake to trust to the highest marks of the surf that may be visible on the beach, and which are probably the vestiges of gales that have occurred within the previous year or two. Any such experience as this is greatly too limited. There is a vast difference between a "*heavy gale*" and a "*great storm,*" such, for example, as that of January 1839, when the wind assumed a force which has since been only once exceeded, in January 1868. And these storms, great as they were, must be regarded as only heavy gales, when we contrast their effects with those of the celebrated hurricane which visited the South of England on 26th November 1703, and which scattered ruin, desolation, and death on every side. Among the records of its effects, I have selected the following facts (from the historical narrative published in 1769,* and Dr. Derham's paper in the London Transactions); and although these extracts may satisfy us that no subsequent storm has equalled this one, they still prove the dreadful violence with which marine works *may possibly* be assailed.

The loss of men and ships in the Royal Navy was 12 vessels, 1611 men, and 524 guns.

Besides these we have the following items :—

Vessels lost at sea	160
Persons drowned in the Thames . . .	24
Wherries lost in the Thames . . .	500

* An Historical Narrative of the great and tremendous Storm which happened Nov. 26, 1703. London, 1769.

Persons killed in London . . .	123
Total number lost at sea	8000
Houses blown down	800
Churches stripped of lead . . .	100
Steeples blown down	7
Windmills destroyed	400
Trees blown down	29,000
Stacks of chimneys blown down in London .	2,000

So violent a tempest as this may well be regarded as almost preternatural, and an engineer could hardly be blamed although his work had been unable to withstand its assaults. Still, we must ever be cautious in giving too much weight to the effects of heavy gales which have but recently occurred, and to which the name of "*storm*" is too often improperly given.

Storms occur indeed but seldom, perhaps not once in ten or twenty years, and very *great* storms are, as we have seen, of still rarer occurrence, whereas hardly a winter passes, in which one or two heavy gales do not take place. The error of confounding the very different indications of gales with the effects due to storms, may be well illustrated by the analogous error in bridge engineering, of assuming as our data the levels of ordinary freshes instead of those of great floods, which, when they do occur, occasion, as every one knows, vastly greater damage.

Level assumed by Mud a measure of the Exposure.—I have elsewhere * referred to another feature which will be found of very considerable value in judging of the exposure of a coast. This is *the level below the surface of low-water at which mud reposes.* It may appear unlikely that the disturbance of the surface of the sea occasioned by storms should be propagated to great depths, but there is no want of evidence on this head.

* Proceedings Royal Soc. Edin. vol. iv. p. 200.

Mr. Airy, the Astronomer Royal, has shown, on theoretical grounds,* that at a depth equal to the length of the wave the motion is $\frac{1}{535}$ of that of the surface, and mentions that heavy ground-swells break in a depth of 100 fathoms. Sir J. Coode found, from under-water examinations made with the diving dress, that the shingle of the Chesil Bank was moved during heavy winter storms at a depth of eight fathoms, and Captain E. R. Calver, R.N.,† has seen waves six or eight feet high change their colour from the abrasion of the bottom after passing into water of seven or eight fathoms. The late Mr. Robert Stevenson, in his paper on the alveus or bed of the German Ocean,‡ says—"The dispersion of fishes, evinced by their disappearance from the fishing grounds in stormy weather, tends to show the disturbance of the ocean at the depth of thirty or forty fathoms. This observation I have frequently had an opportunity of making near the entrance of the Firth of Forth. Numerous proofs of the sea being disturbed to a considerable depth have also occurred since the erection of the Bell Rock Lighthouse, situate upon a sunken rock in the sea twelve miles from Arbroath in Forfarshire. Some drift stones of large dimensions, measuring upwards of thirty cubic feet, or more than two tons weight, have, during storms, been thrown upon the rock from deep water. These large boulder stones are so familiar to the lighthouse keepers at this station as to be by them termed 'travellers.'"

To these may be added a curious example of the action of the waves on the bottom near Burlington Harbour, where a well was sunk which discharged water whenever the tide rose to within 4 feet 2 inches of the level of its mouth, and, "during storms," says Dr. Storer, who gave an account of it in

* Encyclopædia Metropolitana, article Waves.

† The Wave-screen, by E. R. Calver, R.N. London, 1858.

‡ Wernerian Nat. Hist. Soc. Trans. 1820.

1815, "the water flows in waves similar * to the waves of the sea." These phenomena were attributed to the pressure of the tide and waves on an outlet of the spring, which was on good grounds supposed to exist in the bottom of the bay outside.

From these statements it may easily be inferred that *in exposed situations mud cannot repose near the surface.* No one would expect to find a muddy shore confronting an open sea, where the deep water approached closely to the shore, though he would not express surprise at finding such a beach on the borders of a land-locked bay or of a sheltered estuary.† Although the *absence* of mud in any locality proves nothing, because the tide-currents may sweep it away, or the geological formation may not produce it, yet its *presence* seems both a delicate and certain test of the lowest limits to which the disturbance originating at the surface has reached.

German Ocean.—Now, applying such a test to the German Ocean, we find no mud in the immediate neighbourhood of Whalsey, in Zetland, and I admit that no absolute conclusion can be drawn from its absence; but within twenty-five miles we find it in from eighty to ninety fathoms below low-water. In the latitude of Wick it occurs in from sixty to seventy fathoms; in the latitude of Kinnaird-head we have it, on the Norwegian side, in forty to fifty fathoms; in the Moray Firth, abreast of Banffshire, we find it in depths of about thirty-five fathoms; while, as we proceed towards the more sheltered parts of that firth, we find it rise to within twenty fathoms of low-water,

* Phil. Trans., 1815.

† At Allippey, in India, there is a peculiar oily mud, existing in such enormous quantities, that when disturbed by the Monsoons, the *sea itself* becomes a mass of fluid greasy mud, which destroys the waves. See Mr. G. Robertson's description (Roy. Soc. Edin. Proceedings).

and within the Dornoch Firth we find it within sixteen, and close in, under the shelter of the Sutherland shore, we find it in only eight fathoms under the low-water surface. In the latitude of the Firth of Forth it appears in depths of from thirty to forty fathoms; and proceeding up that Firth we have a good illustration of the truth of this mud-test; for, on looking at Fig. 4, which represents the relative level of the muddy bottom on the southern shore, we find it gradually rising nearer to the low-water level, in proportion as the shelter increases, from twenty-two fathoms at Dunbar up to three fathoms off Leith; and were the section carried beyond Queensferry, we should find the mud actually emerging above

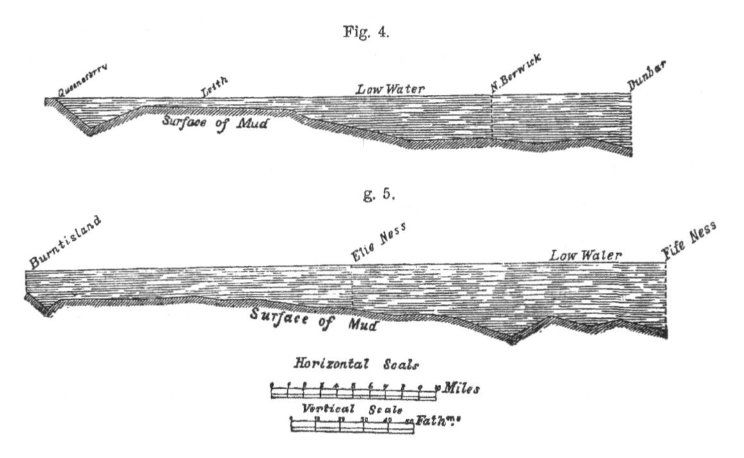

Fig. 4.

g. 5.

the surface at low-water, even although the current gets stronger as we ascend the narrow estuary. Fig. 5 represents a similar section of the northern shore of the Firth of Forth, and here the mud gradually rises from twenty-two fathoms off Fifeness up to *eight* fathoms at Burntisland; whereas near Leith, on the shore opposite (*vide* Fig. 4), it exists as we have stated in *three* fathoms, which accords well with the known fact that the heaviest sea passes on the north side of the Island of Inchkeith, and not on the south.

Leaving this small firth or inlet, which was only adverted to as a proof of the applicability of the test to facts generally recognised, we return to the German Ocean, and observe that towards its southern portion mud is found within about twenty fathoms of the surface, and under lee of the Dogger-bank within fifteen fathoms; and proceeding still farther south, we find it on the coast of Holland at depths of from twelve to sixteen fathoms, and at only eight fathoms at the mouth of the Elbe.

Waves on Dutch Coast.—Now, the violence of the waves upon the shores of the German Ocean certainly decreases in much the same proportion as the rise in the level of the mud, there being a gradual decrease as we come from Shetland and the North of Scotland—where, as will be afterwards shown, wonderful energy is displayed by the sea—to the coasts of Holland, where the waves are much modified. Although it is no doubt true that the flat-bottomed vessels of the Dutch are built purposely for resisting a heavy surf, still the fact of their being able to take the open beach in nearly all weathers along that coast without any protection from harbours, goes far to prove that the waves are much reduced before they reach the Dutch coast. The late Mr. R. Stevenson remarks:*—"On this great range of coast, from Scheveling to the Helder, there is a succession of fishing towns without a single harbour capable of receiving a vessel of almost any description. When the Dutch fisherman therefore arrives upon this coast with a cargo, he allows his vessel to take the ground, when she surges or is driven before the breakers

* Journal of a Trip to Holland: Scots Magazine, 1817.

to high-water mark upon the beach." The comparatively sheltered state of these southern shores is farther corroborated by the practice of the inhabitants in designing the sea walls which protect their coast. Mr. Hyde Clark says * —"On the coast of Zealand they (the Dutch) reckoned 8½ feet as the greatest height to which any wave would be *thrown.*" In short, although in the German Ocean we shall find in any parallel of latitude almost every gradation of depth between the low-water margin of the shores where there is no depth at all, and the maximum sounding in the sea outside, yet *mud nowhere appears to exist in shoal water in any place where there is a heavy sea.*

The same general result may be found on the west coast of the British Isles. While on the west of Ireland mud does not lie nearer the low-water level than from 40 to 60 fathoms, patches may be found on its eastern or more sheltered side, to the north of Dublin, at only 20 fathoms; and half-way up Belfast Lough, where there is good shelter, it may be found at 5 fathoms below the surface.

If, therefore, we find in front of a proposed harbour that mud reposes within a few fathoms of the surface, I believe we have in that fact certain ground for concluding that our works will never be assailed by a very heavy sea.

* Min. Civ. Eng.

CHAPTER III.

GENERATION OF WAVES.

Line of Maximum Exposure—Law of the Ratio of the Square Roots of Distances from the Windward Shore—Coefficient for Heavy Gales—Formulæ for Long and Short Distances—Maximum recorded Heights of Waves in Large Bodies of Water—Partial Action of the Wind—Linear Extent of Gales—Line of Maximum Effective Exposure—Oblique Action of Waves.

Effects of Geographical Configuration of Coast.

Tortuous Channels—Expanding Channels—Propagation of Ground-Swell—Direction of the Coast Line in relation to the Line of Exposure—Lengths, Breadths, and Velocity of Waves.

Line of Maximum Exposure.—In comparing an existing harbour with a proposed one, perhaps the most obvious element is what may be termed the *line of maximum exposure,* or, in other words, the line of greatest *fetch* or *reach* of open sea, which can be easily measured from a chart. But though possessed of this information, the engineer still does not know in what ratio the height of the waves increases in relation to any given increase in the line of exposure.

Law of the Ratio of the Square Roots of the Distances from the Windward Shore.—In 1850 I instituted a series of observations on the Union Canal, on a small fresh-water loch, and also on the Firth of Forth and the Moray Firth, with the view of determining the law of this increase ; and in the Edin. Phil. Journal for 1852, I stated the height of the waves to be most nearly in "*the ratio of the square root of their distances from the wind-*

ward shore," or when h = height of wave, d = distance, and a is a coefficient, varying with the strength of the wind:

$$h = a \sqrt{d}.$$

The truth of this law I have since then had various opportunities of testing.* The accompanying table contains

1 PLACE OF OBSERVATION.	2 Length of Fetch in Miles Nautical.	3 Observed Height of Wave.	4 Height due to Fetch, calculated from Formula, $h = 1.5 \sqrt{d}$.	5 Height due to Fetch, calculated from Formula, $h = 1.5 \sqrt{d} + (2.5 - \sqrt[4]{d})$. *Vide* p. 25.
Scapa Flow	1.0	4.0	1.5	3.0
Firth of Forth	1.3	1.8	1.8	3.2
Granton	2.8	4.0	2.5	3.75
Craignure	3.5	2.0	2.9	3.9
Granton	6.0	4.0	3.7	4.6
Lough Foyle	7.5	4.0	4.1	4.96
Clyde	9.0	4.0	4.5	5.25
Colonsay	9.0	5.0	4.5	5.25
Dysart	10.0	4.2	4.9	5.5
Invergordon	11.0	3.5	5.0	5.7
Lough Foyle	11.0	5.0	5.0	5.7
Glenluce Bay	13.5	5.5	5.6	6.1
Anstruther	24.0	6.5	7.5	7.7
Lake of Geneva, stated by Minard †	30.	8.2	8.2	8.37
Buckie	31.0	7.0	8.4	8.5
,,	38.0	7.0	9.2	9.2
,,	38.0	8.0	9.2	9.2
,,	40.0	8.0	9.55	9.5
Macduff	44.5	8.0	10.02	9.9
,,	45.5	10.0	10.20	10.0
Douglas, Isle of Man, St. George's Channel	65.10	10.12	.12	
Kingstown ‡	114.0	15.0	16.0	15.25
Sunderland, distance measured from Broken Bank	165.0	15.	19.3	18.15
		139.7	153.57	162·68
	Mean .	6.3	6.97	7.39

* It follows from this law that the height of embankments of reservoirs above the water-surface should, *cæteris paribus*, be proportional to the *square roots* of the lengths of water over which the wind acts.

† Cours de Construction des Ouvrages Hydrauliques : Liege, 1852, p. 8.

‡ Some of the extreme waves appeared to be about twice this height, but it is of course very difficult to judge the height of such "*toppling*" waves by the eye.

some of the observations made in 1850-52, as well as later observations on the effects of heavy gales, which could only be made at long intervals of time. Some of these were made for me in various quarters by resident engineers and inspectors of different marine works. For one observation in the Irish Sea I am indebted to Mr. R. Mallet, C.E., and for many others to Mr. Middlemiss, Inspector of Harbour Works. Several are from my own observations, and two of the heights are from estimation by the eye.

Coefficient for Heavy Gales.—Some of these results * have also been laid down in Fig. 6, so as to form a *storm-curve.* Within the limits of the observations, the following very

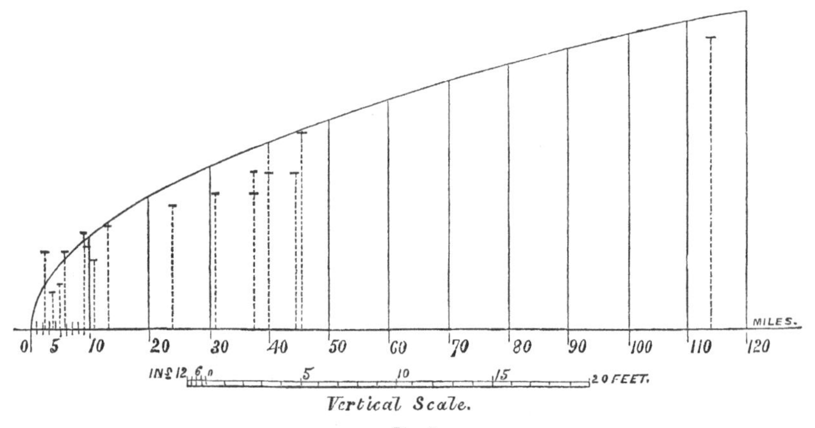

Fig. 6.

simple and easily remembered expression, which is represented by the parabolic curve in the diagram, will indicate pretty nearly the height of waves during heavy gales, at least in seas which *do not greatly differ in depth* from those where the

* The later observations are not given in the woodcut, which was made several years ago.

stations were situated. If h = the height of the wave in *feet* during a strong gale, d = length of exposure in *miles*; then, unless the water is of insufficient depth to allow the waves to be fully formed, or becomes so shallow as to reduce their height after they are formed—

$$h = 1.5\sqrt{d}.$$

The heights calculated according to this formula are given in column 4 of the foregoing Table.

Formula for Short Distances.—For most practical purposes of the engineer, this formula will be found sufficiently accurate. In order, however, to secure strictly comparable results, an anemometer would be required. But it must be observed that in short fetches in narrow lochs or arms of the sea, waves are raised higher during very violent gales than the formula indicates; though it does not appear that such waves go on progressing in the same high ratio for any considerable distance. With the view of rendering the formula more exact for short reaches and violent squalls, the following will be found more suitable:

$$H = 1.5\sqrt{D} + (2.5 - \sqrt[4]{D}).$$

As already mentioned, however, the first easily remembered formula will be found quite sufficient for all ordinary gales and distances, and for bodies of water similar to those where the observations were made. Column 5 of the Table of Observations contains the results given by this last formula.

Mr. Hawkesley, in Min. Civ. Eng.,* published in 1861, says —"In the heaviest gales with which the British Coast was visited, their (the waves) height in yards was represented by the square root of the length in yards of the run of any one wave divided by 40." This testimony is satisfactory as corroborating the law of increase as the square root of the distance, which was given in my formula of 1852, but assigns a coefficient which gives

* Vol. xx. p. 361.

much greater results than my experience warrants. The following Table may be found convenient as a general guide :—

TABLE showing APPROXIMATE HEIGHTS of WAVES due to Lengths of Fetch.

Miles.	Heights.	Miles.	Heights.	Miles.	Heights.	Miles.	Heights.
1 =	3.0	20 =	7.1	39 =	9.4	130 =	17.1
2	3.4	21	7.2	40	9.5	140	17.7
3	3.8	22	7.4	41	9.6	150	18.4
4	4.1	23	7.5	42	9.7	160	19.0
5	4.3	24	7.6	43	9.8	170	19.5
6	4.6	25	7.8	44	9.9	180	20.1
7	4.8	26	7.9	45	10.0	190	20.7
8	5.0	27	8.0	46	10.2	200	21.2
9	5.3	28	8.1	47	10.3	210	21.7
10	5.6	29	8.3	48	10.3	220	22.2
11	5.7	30	8.4	49	10.5	230	22.7
12	5.9	31	8.5	50	10.6	240	23.2
13	6.0	32	8.6	60	11.6	250	23.7
14	6.2	33	8.8	70	12.5	260	24.2
15	6.3	34	8.8	80	13.4	270	24.6
16	6.5	35	8.9	90	14.2	280	25.1
17	6.7	36	9.0	100	15.0	290	25.5
18	6.8	37	9.2	110	15.7	300	26.0
19	7.0	38	9.3	120	16.4		

Maximum recorded Height of Waves in Large Bodies of Water.—The only observations which I have met with for longer fetches than those in the Table are a height of 14 feet 10 inches, as ascertained by the Comte de Marsilli, in 1725,* in the Mediterranean, where the longest possible fetch is about 600 miles. At the harbour of Lybster, Caithness-shire, with the same maximum length of fetch, where observations were made for me during a period of several years, the waves at the shore attained the height of 13½ feet. At Sunderland I found the

* Histoire Phisique de la Mer, par Louis Ferdinand, Comte de Marsilli. Amsterdam, 1725.

waves to be also about 13 feet high at the pier-head; but the height, as in the two former cases, was no doubt reduced by the shallow water near the shore. At Wick, with much the same exposure, waves of about 40 feet have been observed to strike the breakwater. Commander Dayman observed that the highest waves off the Cape of Good Hope were 20 * feet, and Mr. Cockburn Curtis informs us that the gales which produce these rollers extend to from 300 to 600 miles. In the Atlantic Ocean, Dr. Scoresby, when at sea, measured the waves with great care and accuracy on different occasions. He says †—"In the afternoon of this day (4th March 1848) I stood sometimes on the saloon deck or cuddy roof watching the sublime spectacle presented by the turbulent waters. I am not aware that I ever saw the sea more terribly magnificent." Dr. Scoresby then ventured to the port paddle-box. "Here also," says he, "I found at least *one half* of the waves which overtook and passed the ship were far above the level of my eye" (30 feet 3 inches above the level of the sea). "Frequently I observed long *ranges* (not acuminated peaks) extending 100 yards perhaps on one or both sides of the ship, the sea then coming nearly right aft, which rose so high above the visible horizon as to form an angle estimated at 2 or 3 degrees (say $2\frac{1}{2}^{\circ}$) when the distance of the wave-summit was about 100 yards from the observer. This would add nearly 13 feet to the level of the eye, and this measure of elevation was by no means uncommon, occurring, I should think, at least once in half-a-dozen waves. Sometimes peaks of crossing or crests of *breaking* seas would shoot upwards at least 10 or 15 feet higher. The *average wave* was, I believe, fully equal to that of my sight on the paddle-box, or more—that is $\frac{30}{2} = 15$ feet or upward, and the *mean highest waves*, not including the broken

* Min. Civ. Eng. † Brit. Assoc. Rep. 1850.

or acuminated crests, about 43 *feet above the level of the hollow occupied at the moment by the ship.*"*

Partial Action of the Wind.—Dr. Scoresby adds that, "in respect to form we have perpetual modifications and varieties, from the circumstance of the inequality of operation of the *power* by which the waves are raised. Were the wind perfectly uniform in direction and force, and of sufficient continuance, we might have in wide and deep seas waves of perfectly regular formation. But no such equality in the wind exists. It is perpetually changing its direction within certain limits, and its force too, both in the same place and in proximate quarters. Innumerable disturbing influences are therefore in operation, generating the varieties observable in natural seas."

Any one who has watched attentively the forms of waves will confirm this statement. During the continuance of a gale they assume a very irregular appearance, and defy all attempts to trace any individual undulation for a long distance. This irregularity in the action of the wind may be still better seen in a field of ripe corn. At some parts of the moving stalks there appears a concentration to a very near centre, while all around there is but little deflection. In short, while the whole field appears to be acted on by the wind, numerous small patches of greater depression are here and there to be seen, and these depressions quickly disappear, while others are as speedily formed at some distance in advance. Proofs of the same irregularity of action may be found after the gale has subsided, providing it has been strong enough to lay the corn. Instead of finding the whole of the surface depressed, the stalks will be found in some places erect, and at others flattened, and these spots alternate the one with the other, without any regard to regularity.

* Brit. Assoc. Rep. 1850, p. 26.

Linear Extent of Gales.—The reader must not, however, assume that the height will in every case be proportional to the line of exposure. Though this be true of the smaller class of seas, it cannot be extended to large oceans. It is probable that few gales are of sufficient extent to act over such a distance as between Europe and America; and though it did, we may certainly conclude that there is a limit to the height to which the waves can be raised. Those noticed by Scoresby in the Atlantic, which were 43 feet, may perhaps have nearly attained the maximum height for any gale, however great the depth, or however long the distance over which it acts. Though it is generally believed that the Atlantic gales have a rotatory motion, it is still quite consistent with that theory that the undulations should continue to be raised in one direction by the action of the storm for very considerable distances. The late Colonel Reid, in his Development of the Law of Storms, says, at p. 32—"I apprehend that the great undulations raised by the wind in revolving storms are raised along the radii of the whirlwind circle, and roll straight onwards in the direction of tangents to the circle of the whirlwind." Again, at p. 35—"The undulations raised by storms sometimes roll on to a very great distance. . . . I was in Bermuda when the hurricane of 1839 occurred, and distinctly heard the sea breaking loudly against the south shores on the morning of the 9th September, full three days before the storm reached the islands, as recorded in the tables of the state of the weather, kept at the central signal-station. At that time the hurricane was still within the tropics, and distant ten degress of latitude. *As the storm approached, the swell increased, breaking against the southern shores with louder roar and great grandeur*, until the evening of the 12th September, when the whirlwind storm, reaching the Bermudas, set in there." From these facts Colonel Reid draws the following conclusion :—" I think it probable that the heaviest swell proceeding from a storm may be that which is propelled forward in the track which the storm is itself following, as the undulations in this case would be constantly receiving renewed impulses from the storm in its progression. This may account for the unusual degree of grandeur with which the undulations broke against the southern shores of the Bermuda Islands just before the storm set in."

Mr. Redfield* says that on one flank of the Atlantic storms, the direction of the wind coincides with that of the wave-propagation, and he also gives the extent of the following remarkable hurricanes, which I have arranged in a tabular form :—

Storm of	Followed a Straight Course for	Mean Velocity.
23d June 1831	1700 miles	17 miles per hour.
29th Sept. 1830	1800 "	25 " "
10th Aug. 1831	2000 "	13½ " "
Sept. 1804	2200 "	15½ " "
12th Aug. 1835	2200 "	15½ " "
Aug. 1830	3000 "	17 " "
17th Aug. 1827	3000 "	11 " "

To these instances we may add that the westerly gale of December 1862 blew for two and a half days at Edinburgh without almost any variation in direction. It will be found, by any one paying attention to the subject, that gales of as long continuance in one direction as that referred to, which was not specially selected for illustration, are by no means of rare occurrence in Britain.

Line of Maximum Effective Exposure, Oblique Action of

* Journal Franklin Institute, vol. xix. 1837.

Waves.—It does not follow, however, that the line of maximum exposure is in every case the line of maximum effective force of the waves, for this must depend not only on the length of *fetch*, but also on the angle of incidence of the waves on the walls of the harbour. What may be termed the line of *maximum effective exposure* is that which, after being corrected for obliquity of impact, produces the maximum result, and this can only be ascertained from the chart by successive trials. Let x = the greatest force that can assail the pier, h = height of waves which produce (after being corrected for obliquity) the maximum effect, and which are due to the line of maximum effective exposure, $\sin\alpha$ = sine of azimuthal angle formed between the directions of pier and the line of maximum effective exposure. Then, when the force is resolved normal to the line of the pier—

$$x \propto h \sin^2\alpha\,;$$

but if the force be again resolved in the direction of the waves themselves, the expression becomes

$$x \propto h \sin^3\alpha.$$

It should not, however, be forgotten, in connection with this subject, that in some cases there are qualifying elements to which special attention requires to be given. The waves, for example, when approaching the land obliquely, often alter their direction when they get close to the shore, in consequence of a change in the depth, and from this cause they strike more nearly at right angles to the general line of the beach, and thus strike with greater force than the line of maximum effective exposure would lead us to expect.

Reduction of Force due to Oblique Action.—Although experimental observations are still wanted, we are not without practical proof of the reduction of the force of waves where the obstacle lies obliquely to their direction. At the harbour-works of Lybster, in 1851, during the erection of the pier-head, which stands at right angles to the waves, occasional damage took place, and during one gale three stones about a ton each were thrown down, while the wharf wall immediately adjoining, which was parallel to the motion of the waves, was never injured in the slightest degree, although it was of far inferior strength. From the repeated injuries that the pier-head sustained while it was in progress, it was found necessary to connect together the whole of the stones with bolts—a precaution which was not required at the quay wall. The late Mr. James Bremner of Wick, who had much experience in sea works, recommended that piers should be laid out so as to form a horizontal angle of not more than 25° with the heaviest billows ; while Professor Airy, on the other hand, considers that it is safer for the sea to impinge at right angles.

The extraordinary difference between waves acting only at right angles, and others having even a very slight amount of obliquity, has been shown in the most unmistakable manner in the Wick Breakwater, elsewhere referred to, where all attempts to make the work stand, when exactly at right angles to the sea, have hitherto been unsuccessful. Although I know no other work which is exposed to the same class of waves, still the lesson may be useful in other less exposed situations, by directing the attention of the engineer to the necessity of adopting additional precautions. The waves, on entering the bay of Wick, assume a curved form *in plano*, and impinge upon the outer part of the outer kant of the new breakwater at nearly normal incidence. It was found by observation, that while waves coming from the direction of S. by E. struck the outer part at normal incidence, they struck the landward end of the same kant at an angle of incidence of 81°, giving 9°, of obliquity. Other south-easterly seas, which struck the outer part of the outer kant at an angle of incidence of 73° struck the landward part of the same kant at an angle of incidence of 68°, giving 5° of obliquity.

Effects of Geographical Configuration of the Coast.

Narrow Tortuous Channels.—The value of the line of maximum effective exposure varies in certain localities with the geographical configuration of the land. The harbour of Inverary lies at the termination of Loch Fyne, which, from its being narrow and somewhat tortuous, might lead to the supposition that no heavy waves would reach its upper end. I have been informed, however, that waves of very considerable height do strike the pier of Inverary, and are occasionally found troublesome to the small steamer which crosses the St. Catherine's Ferry. From the mountainous character of the country, it is probable that when a strong gale blows, it successively alters its direction with that of the valleys, so as in a great measure to counteract the effects of the *winding* of the loch.

Expanding Channels.—On the other hand, where the channel enlarges, the height of the waves is decreased. Craignure Bay, on the northern shore of the Island of Mull, lies at the eastern end of the Sound of Mull, and has a line of maximum exposure extending for about 25 miles up Loch Lynnhe. As the water is deep, this circumstance would lead us, though the channel is no doubt narrow, to expect from the formula that waves approaching eight feet in height would break upon the shore at Craignure, but I doubt whether they ever attain nearly such a magnitude. The tides may perhaps have some effect in reducing their height; but if we consider the geographical configuration, we shall find that there are other causes to account for the reduction of height. At the southern termination of Loch Lynnhe, where Craignure lies, the channel is bifurcated, one branch leading southwards between the Islands of Lismore and Mull, and the other leading northwards through the Sound of Mull. When, therefore, waves which are generated in the long fetch of Loch Lynnhe reach the southern end of Lismore Island, they lose their height by expanding to the south round Lismore, to the north into the Sound of Mull, and to the west into Craignure Bay. It will be seen from the subjoined register, how small a height the waves attained during the time specified, extending from September till February of the years 1853-54.

Height of Waves at Craignure, when wind blew down Loch Lynnhe.

1853,		Ft.	in.	1863,		Ft.	in.
Sept. 25.	Gale	2	3	Dec. 28.	Moderate ...	0	9
Oct. 6.	Very strong	2	2	,, 30.	Fresh gale...	2	0
,, 11.	Moderate ...	1	0	,, 31.	Do.	1	5
,, 12.	Light.........	0	9	1854,			
Dec. 6.	Light.........	0	5	Jan. 4.	Fresh.........	1	0
,, 13.	Strong	1	4	,, 5.	Strong	1	9
,, 14.	Fresh.........	1	4	,, 6.	Moderate ...	1	0
,, 15.	Moderate ...	0	8	,, 8.	Strong	2	4
,, 20.	Light.........	1	1	,, 9.	Moderate ...	1	4
,, 26.	Fresh.........	0	10	,, 10.	Moderate ...	0	10
,, 27.	Moderate ...	0	6				

Propagation of Ground-swell in Narrow Channels.—At Dunoon, in the Firth of Clyde, the ground-swell, after coming through the narrow passage at the Cumbraes, is still from 7 to 9 feet high. It therefore passes through a neck of only $1\frac{1}{4}$ mile broad, and is propagated through a channel $1\frac{3}{4}$ mile broad for a distance of 14 miles *when there is no wind;* it even reaches Gourock and Greenock after turning through more than a right angle, and after losing height by divergence into several capacious lochs. The ground-swell at the Firth of Forth passes above Queensferry, which is about $\frac{8}{10}$ of a mile broad, and is upwards of 30 miles from the mouth of the Firth.

Direction of the general Coast Line in relation to the Line of Exposure.—In 1857 I issued a series of queries among fishermen and others at various parts of the coast of Scotland, as to the direction from which the heaviest seas come upon the coast. Though there are some apparent anomalies, the general result derived from the statements of nearly 300 fishermen and others is, that at the distance of 1½ mile seaward of the coast line, the heaviest waves come in the direction of the longest fetch, which goes to corroborate the supposition that gales frequently act over large extents of water. On the shore, however, the force is much modified by the angle formed by the coast with the line of maximum exposure. On the east coast it was found that, at about 1½ mile off the shore, the north-east is generally the worst direction; but for that part of the coast which extends from the Tay to Aberdeen, the south-east waves generally break heaviest upon the shore. This arises from the small angle which the north-east bearing makes with the land at this part of the coast. *The most exposed coasts may therefore be regarded,* cæteris paribus, *as those on which the waves generated in the line of maximum exposure come dead-on upon the shore.*

Length and Velocity of Waves.—The longest distance apart, from crest to crest, of the Atlantic waves observed by Scoresby was 790 feet.* His other results are as under: –

* Life of William Scoresby by Dr. Scoresby Jackson. London, 1861, pp. 159 and 324.

Altitude, 43 feet.
Mean distance between waves, 559 feet.
Interval of time between each wave, 16 seconds.
Velocity of each wave per hour, 32½ miles.

The following were observed by Mr. Douglas at the Bishop Rock, on whose authority they are stated :—

8 feet waves,	35 in a mile,	171 feet apart.	8 per min.
15 ,, ,,	5 & 6 do.,	1200 & 1000 ft. do.	5 ,,
20 ,, ,,	3 do.,	2000 feet do.	4 ,,

The late Mr. Mackintosh, lightkeeper at the Calf of Man, in the Irish Sea, informed me that he had, on three different occasions, counted 13½ waves between the Calf of Man and the Chickens Rock. This distance gives about 490 feet as the length of the waves in this comparatively landlocked branch of the ocean.

CHAPTER IV.

FORCE OF THE WAVES.

Force in small bodies of Water—Remarkable destructive Effects at Whalsey Skerries in Shetland—Extraordinary Force at Wick Breakwater—Marine Dynamometer—Formula—Other forms of Dynamometer—Forces indicated by Dynamometer—Relative Force of Summer and Winter Gales—Greatest recorded Force in the Atlantic and German Oceans—Forces exerted at different Levels—Proofs of the Accuracy of Results of Dynamometer—Inadequate Ideas as to Force of Waves—Answer to Objection to Results—Answer to Objection of referring Results to a statical Value—Concentrated Action produced by all Waves in breaking.

SMEATON, when referring to the propriety of using joggles in the masonry of the Eddystone Lighthouse, says—"When we have to do with, and to endeavour to control, those powers of nature that are subject to no calculation, I trust it will be deemed prudent not to omit in such a case anything that can without difficulty be applied, and that would be likely to add to the security." This statement of our greatest marine engineer indicates the propriety of carefully collecting any facts that may help us to a more accurate estimation of those forces which he regarded as being "*subject to no calculation.*" We shall therefore state a few facts which have been recorded of the destructive power of the waves both in small bodies of water and in the open ocean.

Inland Lochs.—At Port Sonachan, in Loch Awe, where the fetch is under 14 miles of fresh water, a stone weighing a quarter of a ton was torn out of the masonry of the landing-slip and overturned. Mr. D. Stevenson, in his *Engineering of North America*, describes the harbours in Lake Erie as reminding him of those on our sea-girt shores, and mentions having seen at the harbour of Buffalo one stone, weighing upwards of half a ton, which had been torn out of its bed, moved several feet, and turned upside down. At the Bishop Rock Lighthouse, a bell was broken from its attachments at the level of 100 feet above the high-water mark during a gale in the winter of 1860,* and at Unst, the most northern of the Zetland Islands, a door was broken open at a height of 195 feet above the sea. To these facts it may be added, that I know, from the testimony of an *eye-witness*, of a block of 50 tons' weight being moved by the sea at Barrahead, one of the Hebrides.

Remarkable Destructive Effects at Whalsey Skerries.—But still more extraordinary effects have been observed at Whalsey, in Zetland, where heavy blocks of rock have been quarried, or broken out of their beds *in situ* on the top of the Bound Skerry, at a great elevation above the sea. Though there are probably few places where the waves are so violent and dangerous as at Whalsey, still it is well for the reader to be able to recognise the characteristic appearances of similar dangerous localities; and to be put on his guard by a description of the place and the phenomena which it presents; *for it must be distinctly understood that in such places the ordinary methods of construction cannot be applied.*

The Bound Skerry is the most eastern of the Shetland group. It consists of quartz rock, forming a part of the gneiss strata, which are here permeated to a considerable extent by "dries" or seams, and, with the exception of a species of lichen that grows on the higher parts, little or no vegetation is to be seen on its surface, although it attains at one point an elevation

* Nautical Magazine, vol. xxxi. p. 262.

of 80 feet above high-water, and about 86 feet above low-water spring tides. The specific gravity of the rock was found to be 2.698, or about 13.3 cubic feet to the ton. The calculations of the weights of the blocks that were moved I have taken, however, at 14 feet to the ton, in order to be fully within the mark. The accompanying sections, Figs. 7, 8, were made with

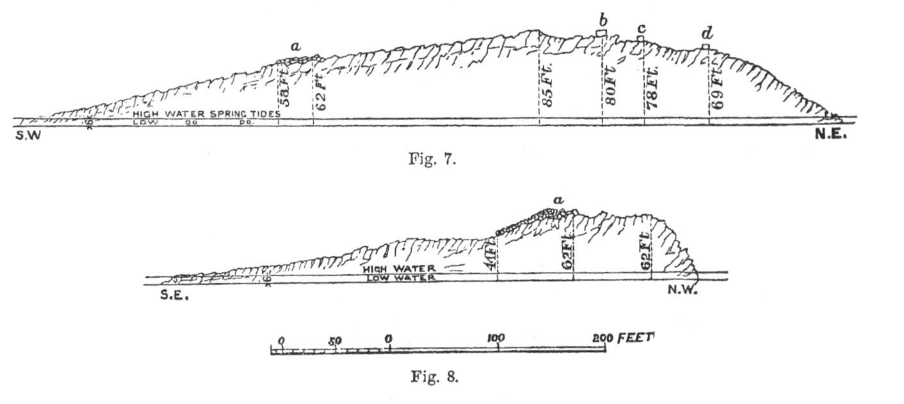

Fig. 7.

Fig. 8.

the spirit-level, and represent elevations of the skerry in the north-eastern and south-western directions, which are the most exposed. It must not, however, be supposed that there is any approach to uniformity of contour, even at places which are very near to each other. The whole island, indeed, forms one of the most rugged and irregular rocks that can well be imagined.

In 1852, when landing for the first time upon this skerry, in order to fix upon the best site for a lighthouse, my attention was speedily attracted by some unmistakable indications of a violent destructive agency which seemed to have been lately at work upon the hard rock of which it consists. These were, the presence of loose blocks of a very large size, which had been detached from the adjoining strata. The only visible agent was the ocean, the unruffled surface of which appeared far below the place where I stood—not less, indeed, than 70 feet, as the levels afterwards proved. Under circumstances so unlikely, it will not appear strange that I did not readily persuade myself that the sea was really the agent of destruction. But, after wandering for an hour or more over the surface of this rugged islet, it was impossible any longer to doubt that the remarkable effects which I had noticed were due to the sea alone. I landed on the Bound Skerry with what I thought tolerably certain and definite conceptions, not hastily adopted, but the result of nearly twenty years' study of the action of the waves at different parts of the coasts of Britain; but I came away with greatly altered views. In order to satisfy myself fully as to the matter, I proceeded to the adjoining islands of Gruna and Brury, where at almost every step similar proofs of violent action presented themselves. At Brury, for example, the ground was covered with large recently moved blocks, at an elevation of 45 feet above high-water.

To return, however, to the Bound Skerry, it may be stated that a considerable portion of the rock which confronts the south-eastern round to the north-eastern seas is in a state of rapid disintegration. On the south-east side, about 370 feet from the low-water mark, and at a height of $62\frac{1}{2}$ feet above its level, there occurs a remarkable beach of angular blocks varying in size from about $9\frac{1}{2}$ tons downwards, and huddled together just as one would have expected to find, had they been elevated only a few feet above the high-water level. This beach of stones appears in the sections, at *a*, Figs. 7 and 8. A little farther seawards was found a detached block of 19.5 tons.

Towards the north-east, at the level of 72 feet above

the sea, in addition to many smaller blocks which had evidently been recently detached, there was one 5½ tons in weight (*b*, Fig. 9). It presented the appearance of

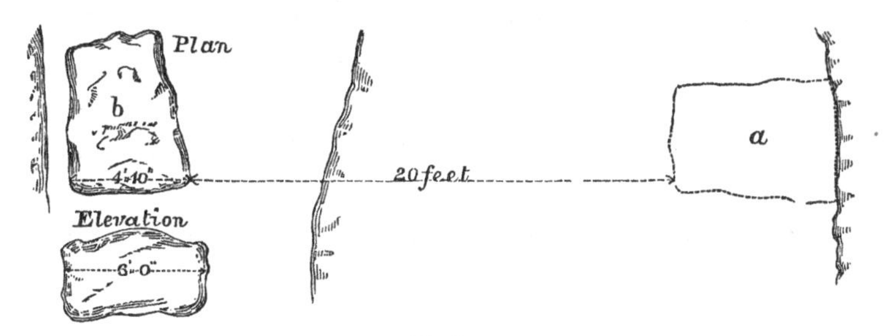

Fig. 9.

recent detachment, having a fresh unweathered look. Within 20 feet of the spot where it lay, there was a comparatively recently-formed void in the rock, which, upon examination and comparison by measurement, was found to suit exactly the detached block. Here, then, was a phenomenon so remarkable as almost to stagger belief—a mass of 5½ tons not only moved but actually quarried from its position *in situ* at a level of 72 feet above high-water spring tides. But higher up still there was another detached rock (Fig. 10), weighing no less than 13½

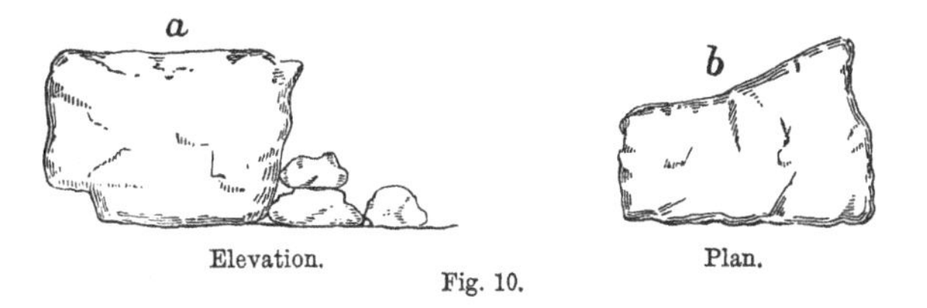

Elevation. Plan.

Fig. 10.

tons, tilted up in a peculiar position, and underneath which numerous angular masses had been wedged, obviously by aqueous action. This great block (Fig. 10) was, however, unlike the one first described, in bearing no traces of *recent* displacement. Though covered with lichen, and apparently long undisturbed, yet there can be no doubt that it too had been separated from the parent cliff, and been tilted up into the position which it now occupies by no other agency than that of the sea, the high-water margin of which is 74 feet below it.

Some persons have suggested that these effects, which must have been the result of an enormous mechanical power, might have been produced by artificial means or by lightning. These explanations, however, are altogether untenable ; for, without touching upon other objections that might be urged against such hypotheses, the wide-spread and frequent recurrence of similar, though not quite so remarkable appearances on the adjoining islands, furnishes, to any one who will spend a day in exploring the rocks at Whalsey, abundant proofs that such explanations are quite insufficient. In order, however, to remove as far as possible any doubts that may exist as to the waves having been the sole cause of these destructive effects, I shall describe a block which was discovered on a lower rock on the south-east side of the skerry. This mass of 7½ tons, represented in Fig. 11, rested upon rugged

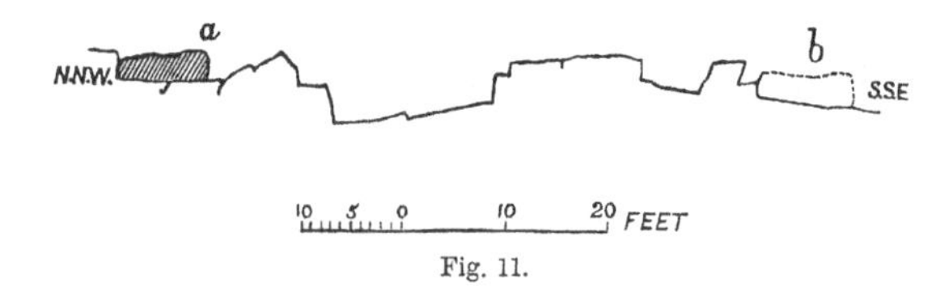

Fig. 11.

peaks of rock at the level of 20 feet above the sea. That it had been very recently detached, no one who saw it could for a moment have a doubt. It was wedged against a high ledge of the skerry ; and that it had been driven against this obstacle with great violence was proved by the fact that both the block itself and the rock *in situ* were smashed and splintered at the point of contact. All its margins were scored and broken, just as might have been expected from its successive contact with the different ledges of rock over which it had been moved. The original position *in situ, b*, from which it had been detached, is shown by dotted lines. It was easy to trace the course which it had taken, as there were everywhere visible unmistakable marks of the violence with which it had been driven along its rocky path, which was of the most irregular description, presenting abrupt and almost vertical faces of from 2 to 7 feet in height. At the distance of 11 feet from the block lay loose fragments, one of which required three men to lift, and which was found to fit exactly into a void in the large block. The original position *in situ* was 73 feet distant in a south-south-east direction, and the void was found to agree in every respect with the travelled block, both in shape and dimensions. The gale which detached this mass was thus proved to have been from the south-south-east, and the maximum length of "*fetch*" corresponding to this direction is about 500 miles.

The fact that a block of nearly 8 tons had been torn up and driven before the waves at the level of 20 feet above the sea for a distance of 73 feet over such rugged ledges, is certainly very remarkable, though it cannot compete with the instances formerly mentioned, where the masses were from 6 to 13 tons in weight, and had been forced from their original beds at places which are from 70 to 75 feet above the sea. This less remarkable fact is adduced merely with the view of supplying a link in the chain of evidence which connects the sea with the movement of the larger blocks at the higher level. Mr. D. Stevenson has since found similar, though not such remarkable appearances, on other islands on the north-east of

Shetland, which led him to the conclusion that these violent effects are generally characteristic of those seas. My last visit to Whalsey having been fortunately made in company with the late Sir Roderick Murchison, I very willingly add the testimony of so distinguished a geologist. He says *—"Mr. Stevenson here" (at Bound Skerry) "called my attention to the manifest proofs of the remarkable power of the sea-waves when lashing upon this exposed spot in great storms. The seaward or north-eastern face of the gneissose rocks sloping upwards, presents the most chaotic aspect, being covered with clusters of large angular blocks, one of the largest of these being at nearly 70 feet above the sea. Now, all of them have been torn out of their beds, and most of them moved up-hill for a considerable number of feet, to within a few yards of the base of the new lighthouse. For my own part, I was at first incredulous as to the mode of producing what my lamented friend, Leopold von Buch, would have called a true 'Felsen Meer;' but when Mr. Stevenson brought the data before me, it was quite evident that the sea had done it all. Thus, an inhabitant pointed out some of the chief blocks, several of them many tons' weight, which, in a great storm some years back, had been moved upwards on the incline 15 to 20 feet, to heights of 50 feet above the sea. These, in their upward translation, had scored the rocks over which they passed, just as the stones held in a glacier groove and scratch in their descent; and the freshness of the markings was quite striking. Not trusting to histories of the past, and for a moment doubting even the clear evidence offered by the scoring of the rugged subjacent rocks, I interrogated an intelligent under-officer of the lighthouse, who had been two years on the spot, and ascertained that, even in the preceding winter, and when

* Quarterly Journal of the Geo. Soc. of London, 1859, vol. xv. p. 392.

the new lighthouse was in course of construction, a huge mass of stone near the sea-level, of which he showed the very bed out of which it had been lifted, had been wrenched out of it, and moved up an incline of 10° or 12° to a distance of 16 feet! and with this proof all scepticism vanished."

Extraordinary Force of the Sea at Wick Breakwater.—When we wish to ascertain what is the greatest feat that has been achieved by the waves, we naturally look to the ravages which are to be discovered in the rocky cliffs which confront the ocean. We should never expect to find examples of the development of the *greatest* force against the masonry of those artificial works which form our ports and harbours. The enormous extent, and endless variety of exposure, of the shores of Britain, as compared with those of the few piers or breakwaters erected here and there along the line of coast, make it to the last degree improbable that the maximum results should be found anywhere else than among the rocks *in situ* on the shore. Accordingly the examples of the most violent wave-action which have just been mentioned, and which were all that were given in the first edition of this book, are cases of the destruction or movement of dislocated natural rocks. This, however, no longer holds true. The most startling example now on record is that of an artificial work.

The harbour works at Wick, which have been nine years in progress, were commenced in 1863, and consisted of blocks of from 5 to 10 tons, set on edge, first built above high-water neap tides with hydraulic lime, then with Roman, and latterly with Portland, cement. Plate No. XI. shows the position and depth of water in which the breakwater is built. In October 1864, 300 feet of the contractor's staging were carried away; and greenheart was afterwards substituted for memel piles. The depth under low-water springs in which the first portion of the wall was founded was 12 feet, in conformity with universal practice; but 18 feet was afterwards adopted, which was a fortunate precaution, for in 1868 the rubble was washed down to 15 feet below low-water, and serious damage occurred to a part of the superstructure. In 1870 a length of 380 feet (about $\frac{1}{3}$ of the whole) was destroyed. In February 1872, after the superstructure had been rebuilt solid, with Portland cement, a new species of damage took place, the face-stones being in many places shattered by the sea, which is all the more remarkable from the fact that the blocks were of the same density as granite, and of a strength three times greater than that of Craigleith stone—a phenomenon, indeed, unparalleled in the history of sea works. Lastly, in December 1872, the greatest proof of force was manifested, and is thus given in the words of a report by Messrs. Stevenson:—"The (seaward) end of the work, as has been explained, was protected by a mass of cement rubble work. It was composed of three courses of large blocks of 80 to 100 tons, which were deposited as a foundation (in a trench made) in the rubble. Above this foundation there were three courses of large stones carefully set in cement, and the whole was surmounted by a large monolith of cement rubble measuring about 26 feet by 45 feet, by 11 feet in thickness, weighing upwards of 800 tons. This block was built *in situ*. As a further precaution, iron rods, 3½ inches diameter, were fixed in the uppermost of the foundation courses of cement rubble. These rods were carried through the courses of stone work by holes cut in the stone, and were finally embedded in the monolithic mass which formed the upper portion of the pier. The arrangements we have described will perhaps be best understood by the accompanying sketch, Plate XI. Fig. 2. Incredible as it may seem, this huge mass succumbed to the force of the waves, and Mr. M'Donald, the resident engineer, actually saw it from the adjacent cliff being gradually "*slewed*" round by successive strokes until it was finally removed, and deposited inside of the pier. It was not for some days after that any examination could be made of this singular phenomenon, but the result of the examination only gave rise to increased amazement at the feat which the waves had achieved. It was found on examination by diving that the 800-ton monolith forming the upper portion of the pier, which the resident engineer had seen in the act of being washed away, had carried with it the whole of the lower courses which were attached to it by the iron bolts, and that this enormous mass, weighing not less than 1350 tons, had been removed *en masse,* and was resting *entire* on the rubble at the side of the pier, having sustained no damage but a slight fracture at the edges. A further examination also disclosed the fact that the lower or foundation course of 80-ton blocks, which were laid on the rubble, retained their positions unmoved. The second course of cement blocks, on which the 1350 tons rested, had been swept off after being relieved from the superincumbent weight, and some of them were found entire near the head of the breakwater. The removal of this protection left the end of the work open, and the storm, which continued to rage for some days after the destruction of the cement rubble defence, carried away about 150 feet of the masonry (1-7th of the whole), which had been built solid and set in cement. The same remarkable feature of former damage was strikingly apparent in the last damage—*the foundations, even to the outer extremity of the work, remaining uninjured.*"

Plate XII. is a photographic view of the breaking waves, taken after the strength of the storm had passed, and for the use of which I am indebted to the kindness of Mr.

A. Johnston of Wick, by whom the original photograph was made.

Marine Dynamometer.—The value and importance of ascertaining, by direct experiments, the actual force of the waves expressed in pounds per square foot, or some other measure, either statical or dynamical, will readily be admitted. It will, however, require many years' observations before we can expect to have certain information on such a subject, or be enabled to apply the results with confidence in determining the safe limits of construction for marine works. With a view to forward the investigation, the results may be given of some observations which commenced in 1842, at Little Ross Island, off Kirkcudbrightshire, and a detailed account of some of which will be found in the *Edinburgh Transactions.** These observations were made with the marine dynamometer, a simple self-registering instrument which I designed for the purpose.

As there is no contest to which the old proverb "*fas est ab hoste doceri*" is more applicable than in opposing the surges of the ocean, it may be proper to give such a description of the dynamometer as will enable any one to have it made.

D E F D is a cast-iron cylinder, which is firmly bolted at the projecting flanges, G, to the rock where the experiments are to be made. This cylinder has a circular flange at D. L is a door which is opened when the observation is to be read off. A is a circular disc on which the waves impinge. Fastened to the disc are four guide rods B, which pass through a circular plate C (which is screwed down to the flange D), and also through holes in the bottom of the cylinder E F. Within the cylinder there is attached to the plate C a very strong steel spring, to the other or free end of

* Trans. Roy. Soc. Edin., vol. xvi. part i. p. 84.

which is fastened the small circular plate K, which again is secured to the guide rods B. There are also rings of leather, T,

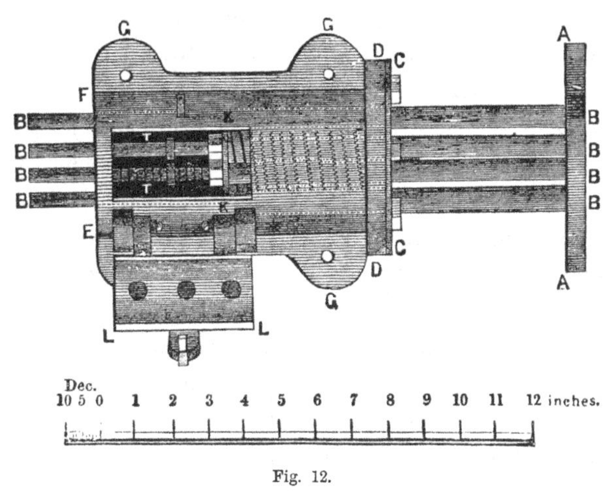

Fig. 12.

which slide on the guide rods, and serve as indices for registering how far the rods have been pushed through the holes in the bottom EF, or, in other words, how far the spring has been drawn out by the action of the waves against the disc A.

The following formula will be found convenient in the graduation of the instrument :—

W = weight stated in tons, which is found by experiment to produce a given amount of yielding of the spring. D = number of inches yielded by the spring with weight W. a = area of the disc in square feet, d = the length in inches on the proposed scale, corresponding to a force of one ton per square foot acting on the disc.

$$d = \frac{D\,a}{W}$$

The different discs employed in the observations referred

to, were from 3 to 9 inches diameter, but generally 6 inches, and the strength of the springs varied from about 10 lbs. to about 50 lbs. for every $\frac{1}{8}$ inch of elongation, and the instruments varied in length from 14 inches to 3 feet. Their respective indications were afterwards reduced to a value per square foot. The instrument was generally placed so as to be immersed at about $\frac{3}{4}$ths tide, and in such situations as would afford a considerable depth of water. It is not desirable to select a site on a much lower level, as it has not unfrequently happened, during a gale, that for days together no one could approach the dynamometer to read off the result and readjust the indices to zero. It must, at the same time, be remarked, that it is in most situations almost impossible to receive the force unimpaired, as the waves are so often more or less broken by hidden rocks or shoal ground before they reach the instrument.

Other forms of Dynamometer.—The instrument which has been described is probably the most convenient that can be adopted, but I have referred in the *Edinburgh Transactions* to other methods which might in some cases be found more suitable.* For example, the impulse of each wave could be noted at such a situation as the Bell Rock or Eddystone Lighthouse by conducting a column of water or air into the interior of the tower. The force of each wave as it struck the building would thus at once be shown either by the rise of the water column, or, if air were used, by means of an attached gauge which would show the same result in atmospheres by compression. The indications of any kind of dynamometer could also be transmitted by wires to some convenient distance, and thus the impulse of every wave could be separately recorded.

Forces indicated by the Dynamometer.—With instruments

* Trans. Roy. Soc. Edin. vol. xvi. part i.

of the kind shown in Fig. 12, the following series of observations, commencing in 1843, were made on the Atlantic, at the Skerryvore, and neighbouring rocks, lying off the island of Tyree, Argyllshire; and in 1844 a series of similar observations was begun on the German Ocean, at the Bell Rock.

Referring for more full information to the tables of observations which are given in the *Edinburgh Transactions*, it will be sufficient here to state generally the results obtained, which were as follows: only premising that the values refer to areas of limited extent, and are applicable therefore only to the *piecemeal* destruction of masonry, and must not be held as applicable to large surfaces of masonry.

Relative Force of Summer and Winter Gales.—In the *Atlantic Ocean*, at the Skerryvore rocks, and at the neighbouring island of Tyree, the average of the results that were registered for five of the summer months during the years 1843 and 1844 was 611 lbs. per square foot = 0.27 tons. The average results for six of the winter months (1843 and 1844) was 2086 lbs. = 0.93 tons per square foot, or more than *three times as great as in the summer months.*

Greatest recorded Forces in the Atlantic and German Ocean.—The *greatest result* obtained at Skerryvore was during the heavy westerly gale of 29th March 1845, when a force of 6083 lbs., or nearly three tons, per square foot, on the surface exposed was registered. The next highest was 5323 lbs.

In the *German Ocean*, the greatest result obtained at the Bell Rock on the surface exposed was at the rate of 3013 lbs. per square foot. But subsequent and much more extended observations at Dunbar, in the county of East Lothian, gave 3½ tons; while, at the harbour works of Buckie, on the coast of Banffshire, the highest result of observations, extending over a period of several years, was three tons per square foot.

Forces exerted at different Levels.—An exposed part of the Skerryvore rock was also chosen, on which two instruments were fixed, the one (No. I.) several feet lower, and about 40 feet seaward of the other (No. II.) It was observed that about half-flood the force of the waves was a good deal expended before they reached the spot where No. I. was

Date.	Remarks.	No. of Instrument.	Pressure in lbs. per Square Foot.
1845,			
Jan. 7.	Heavy Sea.	I.	1714
" "	"	II.	4182
" 12.	Very heavy swell.	I.	2856
" "	" "	II.	5032
" 16.	Heavy ground-swell.	I.	2856
" "	" "	II.	4752
" 22.	A good deal of sea.	I.	2856
" "	" "	II.	5323
" 28.	Heavy ground-swell.	I.	2627
" "	" "	II.	4562
Feb. 5.	Fresh gales.	I.	856
" "	"	II.	3042
" 21.	"	I.	1827
" "	"	II.	3422
" 24.	Fresh breezes.	I.	1256
" "	"	II.	3802
March 9.	Ground-swell.	I.	1256
" "	Waves supposed about 10 feet high.	II.	3041
" 11.	Short sea.	I.	1028
" 24.	Heavy sea.	I.	2281
" "	Waves supposed about 20 feet high.	II.	4562
" 26.	Swell.	I.	1256
" "	Waves about 6 feet high.	II.	3041
" 29.	Strong gale, with heavy sea, the highest waves supposed 20 feet high, and the spray rose about 70 feet.	I.	2856
		II.	6083

placed, from there being so little water on the rocks outside. Whereas, when the tide was higher, the waves were, from the greater depth of water, not so much broken when they reached No. II. The results show generally about *twice* the force at No. II. as at No. I.; a fact which proves how important it would be to ascertain the relative forces of the waves at different levels upon our breakwaters and other sea works.

The observations at Dunbar and Buckie prove that the sea may exert a force so great as 3½ *tons over the limited extent of surface presented by the discs,* and that the force varies much with the level at which the instruments are fixed. These results are given, however, not as ultimate data for calculation, but simply as determining the fact that the sea has been known to exert a force equivalent to a pressure of 3½ tons per square foot, *however much more.* There can be no doubt that results higher than these may yet be obtained. Were the observations sufficiently multiplied, we should soon obtain data which would be highly useful in practice, as they would, by reference to existing sea works, show what sizes of stones were necessary for resisting any given force that was indicated by the dynamometer.

Proofs of the Accuracy of the Results of the Dynamometer.—The greatness of the forces recorded by the dynamometer has led some to express doubts as to the accuracy of the results. This is not to be wondered at, for prior to these observations very erroneous ideas were entertained of the impulsive force of the sea. Sir Samuel Brown, for example, in his arguments for adopting bronze lighthouses* takes the force "on each cylindrical foot column," at only 80 lbs. Captain Taylor, in

* Description of a Bronze or Cast-iron Columnal Lighthouse designed for the Wolf Rock or the Skerryvore, by Sam. Brown, Esq., R.N., K.H.: Edinburgh, 1836, p. 14.

advocating his proposed plan for floating harbours of refuge,* allows a pressure of 144 lbs. per square foot, and Minard seems to calculate on only 70 lbs. per square foot.

The doubts that have been expressed were based on the assumption that the action of the waves is the same as the impact of a hard body; and on the objection to expressing a dynamical force by a statical value. Three classes of phenomena, essentially different from each other, may be referred to as proofs that if the indications of the dynamometer do not represent the force actually exerted, the error must be in defect and certainly not in excess. These are—1*st*, The displacement of heavy bodies, proofs of which have already been given; 2*d*, the elevation of spray; and 3*d*, the fracture of materials of known strength. The elevation of the spray, at the Eddystone, the Bishops, and the Bell Rock Lighthouses, is well known. In November 1827, during a ground-swell, *without wind*, the water rose to the gilded ball on the top of the Bell Rock lantern, which is 117 feet above the rock, and as the tide on that day rose 11 feet on the tower, this leaves 106 feet as the height of elevation. On the same day a ladder was broken from its attachment to the balcony at an elevation of 86 feet, and washed round to the other side. It therefore follows that there is a force in action at the foot of the Bell Rock tower competent during ground-swells, when *there is no wind*, to project a column of water to the height of 106 feet, which, according to the laws of hydrodynamics, is due to a pressure of very nearly 3 tons per square foot,—whereas the greatest force recorded by the dynamometer at this place was only 1½ ton.

Beams of Memel timber, called Booms (*vide* Chap. VIII.),

* Plans for the Formation of Harbours of Refuge, by Captain T. N. Taylor, R.N., C.B.: Plymouth, 1840, p. 7.

which are used at Hynish Harbour, Argyllshire, for keeping the sea out of a small tide basin, have been very frequently broken by the waves. They were 20 feet long between the supports, and 12 inches square. Within six years after the harbour was completed, and at different dates, seven of them were broken, though they were of perfectly sound quality. Each of these logs would resist fracture though uniformly loaded with a weight of 30 tons, so that the sea at Hynish must, on seven different occasions, have exerted a force on each boom which may justly be compared to a dead weight of 30 tons uniformly distributed over the logs. Dynamometric observations were made at Hynish for a considerable time, and the highest result recorded at that place was $1\frac{1}{8}$ ton per square foot, whilst the pressure required to break the booms must have been at least $1\frac{1}{2}$ ton per foot of surface exposed. At Pulteneytown Harbour works, which will be afterwards referred to, both Memel and Greenheart logs, placed vertically in the sea, were found altogether insufficient for the contractor's staging, having been invariably broken near the level of high water. For the satisfaction, however, of any who may still have doubts as to the action of this instrument, the following Table is added, which contains simultaneous observations made at Skerryvore with three dynamometers, having not only discs of very different areas but springs of very different powers, and yet the results are almost identical :—

TABLE of OBSERVATIONS made at Skerryvore Rocks, and also at the Island of Tyrii, with three Dynamometers, having springs of different strength, and discs of different sizes, fixed parallel and close to each other :—

Dates.	Lbs. to a Square Foot.	Dates.	Lbs. to a Square Foot.	Dates.	Lbs. to a Square Foot.
1844,		1844,		1844,	
Jan. 16	428	Mar. 4	3369	April 16	642
,, ,,	427	,, ,,	3427	,, ,,	481
,, 28	3422*	,, 7	1069	,, 17	800
,, ,,	2285*	,, ,,	963	,, ,,	856
,, ,,	3313	,, ,,	913	,, ,,	962
Feb. 2	429	,, 10	1925	,, 18	571
,, ,,	457	,, ,,	1925	,, ,,	481
,, 3	429	,, ,,	1713	,, 19	800
,, ,,	457	,, 11	535	,, ,,	535
,, 13	214	,, ,,	481	,, ,,	481
,, ,,	228	,, ,,	456	,, 22	913
,, 15	321	,, 12	3316	,, ,,	428
,, ,,	280	,, ,,	4011	,, ,,	962
,, ,,	321	,, ,,	2970	,, 24	1942
,, 16	428	,, 13	1142	,, ,,	1604
,, ,,	402	,, ,,	1283	,, ,,	1370
,, ,,	343	,, ,,	1283	,, 25	1283
,, 24	1284	April 10	457	,, ,,	343
,, ,,	1364	,, ,,	428	,, ,,	321
,, ,,	685	,, ,,	481	,, 27	457
,, 26	2032	,, 11	800	,, ,,	481
,, ,,	2086	,, 12	343	,, ,,	Night tide 800
,, ,,	399	,, ,,	321	,, ,,	642
,, 27	321	,, 14	571	,, 30	229
,, ,,	321	,, ,,	535	,, ,,	241
,, ,,	342	,, 16	571		
Mar. 4	3316				

Note.—The two marked thus * were too weak, as the leathers were found flattened, and one of the instruments was broken, and was not repaired till the 15th February.

The means of the above nine observations, which were made with only two instruments, are 433 *lbs. and* 415 *lbs. respectively.*

The means of the above eighteen observations, which were made with three instruments, are 1247 *lbs.*, 1183 *lbs.*, *and* 1000 *lbs. respectively.*

Answer to the Objection that the Action of the Waves is the same as the Impact of a Hard Body.—Now the same force, supposing the waves to act like the impact of a hard body, would, in the marine dynamometer, assume very different statical values, according to the spaces in which that force was expended or developed, so that with the same force of impact, the indications of a weak spring would be less than that of a stronger. This appears from the annexed Table, which contains results of a few experiments in which the springs were tested dynamically by the impact of a cannon-ball dropped from a given height upon the disc of each instrument, which was fixed vertically in a framework of timber, with the disc uppermost. It will be seen from the Table, as was to have been expected, that within the limits of the experiments there was for each spring a different ratio between the value registered by the leather index and the calculated momentum of the falling body. If the waves acted by a sudden finite impact like the cannon-ball, we should not have found such harmony between the results of springs of different strength as appears in the Table of Observations at Skerryvore. *The action of a wave, therefore, is not momentary, as in the case of a solid, but is continuous during the period that the disc is immersed in the passing wave.* In short, to make the cases analogous, a continuous succession of cannon-balls should fall on the disc.

TABLE representing Experiments made on the impact of a Cannon-Ball upon Dynamometers having springs of different strengths.

Large Dynamometer. Strength of spring 462.24 lbs. per inch of elongation. Falling weight 32.5 lbs.					
1 Height fallen through by Cannon-Ball, in Feet, or h.	2 Spring elongated in Inches. e.	3 Calculated Velocity at impact in feet per second. $v = \sqrt{h + e}$.	4 Calculated Momentum.	5 Registered Pressure.	6 Registered Pressure. ÷ Momentum.
0.5	0.875	5.67	184.3	404.5	2.195
1.0	1.25	8.02	260.7	577.8	2.216
1.5	1.5	9.83	316.5	693.4	2.191
2.0	1.685	11.35	368.9	779.4	2.113
				Mean	2.179
Small Dynamometer. Strength of spring 156 lbs. per inch of elongation. Falling weight 32.5 lbs.					
1	2	3	4	5	6
0.5	1.5	5.67	184.3	234	1.270
1.0	2.0	8.02	260.7	312	1.197
				Mean	1.233

Hence it follows that within the limits of the experiments the momentum with the strong spring = . . $\dfrac{\text{Registered Pressure}}{2.179}$

Whereas with the weak spring the momentum = . $\dfrac{\text{Registered Pressure}}{1.233}$

Answer to the Objection to referring the Results of a Dynamical Force to a Statical Value.—The objection which has been raised against any statical valuation of the dynamical effect of the waves falls to the ground when we remember that in all sea works we oppose the dynamical action of the sea by the *dead weight or inertia* of the masonry, so that the dynamometer furnishes exactly the kind of information which the engineer requires.

Concentrated Action produced by all breaking Waves.—Although the height of a jet of water will be increased if the obstacle be of a converging form, yet any plane vertical barrier will produce a high jet, for we see it in every sea-wall, and with all conceivable configurations of bottom. The phenomenon is indeed far too common to admit the supposition of its being occasioned by any re-entrant angle in the bottom, but is probably due to the manner in which a breaking wave collapses or curls over upon itself. From observations which I repeatedly made on the shores of the Mediterranean, at a place where the beach, which was gravelly, presented a uniform profile, the waves in breaking were found to converge very symmetrically, and although there was no obstacle but the beach, the spray was invariably raised much higher than the level of the crest of the unbroken wave. This peculiar change of form, which is common to all breaking waves, destroys the parallelism which may have previously existed among any of the moving filaments of fluid, and converges them towards a horizontal axial line, so that the particles at and near such line are driven upwards not only with their own original velocity, but with an increased velocity due to the proportion subsisting between the number of particles that are raised, and the greater number of particles that are finally stopped.

Some measurements of the height of spray against sea-walls will be found in a subsequent chapter.

The following Table, by the late Professor Rankine,* gives examples of heights in feet due to velocities in feet per second as computed by the equation—

$$\text{Height in feet} = v^2 \div 64.4.$$

* A Manual of Civil Engineering, by W. J. Macquorn Rankine, p. 676: Lond. 1862.

It is exact for latitude 54¾°, and near enough to exactness for practical purposes in all latitudes.

v.	Height.	*v.*	Height.	*v.*	Height.	*v.*	Height.	*v.*	Height.
1	.015528	17	4.4875	32·2	16.100	48	35.776	76	89.688
2	.062111	18	5.0310	33	16.910	49	37.282	78	94.471
3	.13975	19	5.6055	34	17.950	50	38.819	80	99.377
4	.24844	20	6.2111	35	19.021	52	41.987	82	104.41
5	.38819	21	6.8477	36	20.124	54	45.279	84	109.56
6	.55900	22	7.5153	37	21.257	56	48.695	86	114.84
7	.76086	23	8.2141	38	22.422	58	52.235	88	120.25
8	.99377	24	8.9439	39	23.618	60	55.900	90	125.77
9	1.2577	25	9.7048	40	24.844	62	59.688	92	131.43
10	1.5528	26	10.497	41	26.102	64	63.601	94	137.20
11	1.8789	27	11.320	42	27.391	64·4	64.400	96	143.10
12	2.2360	28	12.174	43	28.711	66	67.639	98	149.13
13	2.6241	29	13.059	44	30.062	68	71.800	100	155.28
14	3.0434	30	13.975	45	31.444	70	76.086		
15	3.4937	31	14.922	46	32.857	72	80.496		
16	3.9751	32	15.900	47	34.301	74	85.029		

CHAPTER V.

CONDITIONS WHICH AFFECT THE FORCE OF WAVES.

Tides sometimes act as Breakwaters—Causes of Roosts or Races—Velocity of British Races—Bars of Rivers are miniature Races—Tides sometimes increase Surf on Shore—Time of Tide when Surf is heaviest—Damage by Waves in Deep Water—Unfinished Masonry makes Waves of Translation—Waves on Coasts affected by Tides—Relation between Height of Waves and Depths of Water—Depth regulates Height of Waves—Depth in which Waves break—Height of Waves above Mean Level.

The Tides in some cases act as Breakwaters to the Shore.—At some parts of the coast the tides cause waves of an unusually dangerous character, while at others they are found to *run down* the sea. If a harbour work were situated in a *race* or rapid tide-way—such, for example, as those called "roosts" in Orkney and Shetland—the masonry would be exposed to the action of a very trying and dangerous high-cresting sea. As an instance, we may refer to Portpatrick in Wigtownshire, where the violence of the waves is, to a great extent, due to the rapidity of the tides. If, on the other hand, the race or roost runs in such a direction as to be *entirely outside of the harbour and at some distance off*, it will, while it lasts, have a decided tendency to shelter the works, by acting as a breakwater. It was proved by observations made specially for the purpose at Sumburgh Head in Shetland during a south-westerly storm, that so long as the Sumburgh Roost (one of the most formidable in those seas, and more than 3 miles in width) was cresting and breaking heavily, there was comparatively little surf on the shore; but no sooner did the roost disappear towards high water than a heavy sea rolled towards the land, rising on the cliffs to a great height.

The lightkeeper at Sumburgh Head, in a letter to me, says, "We had a very severe gale from the south-west yesterday, and being the first gale we have had from that quarter since you were here, I paid particular attention to the state of the sea in the West Voe through the day. By daylight in the morning it was blowing very hard, with a most terribly heavy sea *rolling into the West Voe and breaking over the top of the banks, while low-water lasted.* But with regard to what you said to me about the tide in the 'roost' acting as a breakwater to the Voe, your opinion is right, for during the last hours of flood* and the first two hours of ebb tide in particular, *a small boat could have gone till within a few yards of the roost between the Lighthouse and the Horse Island, although the sea was still in the same raging state between the roost and as far as the eye could reach towards Fair Isle and away to the west.*" Here, then, is very satisfactory evidence, that the heavy waves were so much reduced in height by breaking in deep water (it is believed not less than about 40 fathoms), that when they reached the shore they were nearly harmless. The modifying and intensifying effects of tide-currents on waves seem to have been entirely overlooked in the discussions regarding the merits of vertical and sloping walls; a subject which will be referred to in another section.

Causes of Roosts or Races of the Tide.—The opinion expressed by a writer in the *Edinburgh Philosophical Journal*—that the cause of races or roosts is merely the meeting of two rapid currents, seems to be erroneous; neither does it appear

* The current turns one hour and a half before high water on the shore.

possible that they are occasioned only by the projection of rocks from the bottom of the sea, as many sailors suppose.

From careful inquiries, as well as from actual personal experience of such dangerous breaking waters as the Boar of Duncansby, and the Merry Men of Mey in the Pentland Firth, it appears that the true cause is the *swell of the ocean encountering an opposing tidal current.* Two rapid tides may meet each other without any dangerous effects, if there be no ground-swell, yet, if they join together in a rough sea, as in coming round the islands of Stroma or Swona in the Pentland Firth, during ground-swells, the effect of their union being to increase the current, highly dangerous waves will be produced. The meeting of the currents, therefore, though not the *cause* of the waves, is nevertheless sure to increase their height, and to make them break. The races which occur in open seas—as, for instance, off headlands and turning points of the coast—are certain portions of those seas in which, with a ground-swell, the waves *break* to a greater or less extent, although the water may be very deep, and there may be no wind at the time. At all such places it will be found that there are rapid tides, and that the breaking waves are produced when the tide runs against a ground-swell. The roosts on the west coast of Orkney or of the Pentland Firth, for example, are worst with *ebb* tides and *westerly* swells, because the Atlantic swell and current of ebb are opposed. Those again on the east coast are worst with *flood* tides and *easterly* swells, because the swell from the German Ocean and current of flood are opposed. Thus at the east end of the Pentland Firth the Boar of Duncansby is well known to rage with easterly swells and a flood tide; whereas, at the west end of the same firth, the Merry Men of Mey are worst with ebb tide and a westerly swell, at which time no boat can enter them without the greatest risk of being swamped. One or two quotations from the "Sailing Directions for the Pentland Firth," which are given in the "North Sea Pilot" for 1857, will give the reader a good idea of the dangers of those troubled waters, and of the peculiar phenomena presented by the roosts.

"Before entering the Pentland Firth, all vessels should be *prepared to batten down*, and the hatches of small vessels ought to be secured *even in the finest weather*, as it is difficult to see what may be going on in the distance, and the transition from smooth water to a broken sea is so sudden, that no time is given for making arrangements." "The *Swilkie* (of Stroma) must be avoided by boats even in the finest weather, for a few years since a boat was drawn down by one of the whirlpools, and all her crew perished." "So distinct is the line of demarcation between the stream and the eddy, that in passing in a steamer from the one into the other, the engines are brought to a standstill and the vessel twisted round with a great velocity." "During the flood-stream in an easterly or south-easterly gale, it is absolutely necessary to keep an offing of 6 to 8 miles abreast of the Pentland Skerries, until the flood-stream has ceased. Three vessels were observed to founder on the 18th August 1848 when attempting, under these circumstances, to run against the flood."

Velocity of different British Races.—It will not appear surprising that such effects are produced when the swiftness of the currents in those northern seas is taken into account. I have collected in a tabular form the velocities of some of the most remarkable races, from which it will be seen that the velocity of one off the Pentland Skerries is nearly *double* that of the well-known "Race of Portland."

Names of Places.	Authorities.	Velocity of Spring Tides in statute miles per hour.
Portland Race	Admiralty Channel Pilot	5.75 to 6.9
Open ocean between Orkney and Zetland	Admiralty North Sea Pilot	5.76
Hoy Sound, Orkney . .	Do. do.	6.90
Holm Sound, do. . . .	Do. do.	6.90
Sumburgh Roost, Zetland .	Do. do.	8.06
Burger Roost, Orkney . .	Do. do.	8.06
Helgate, New York, east current	Prof. H. Mitchell .	8.5
Doris Mor, Argyllshire . .	Captain Bedford, R.N. .	9.22
Gulf of Corrie Vreckan, Argyllshire	Do.	9.83
Roost near Louther, Pentland Firth	Admiralty North Sea Pilot	10.36
Roost near Swona, Pentland Firth	Do. do.	10.36
Roost near Pentland Skerries .	Admiralty Survey .	12.20

Bars at the Mouths of Rivers form miniature Races.—The dangerous surf which exists at the mouths of some rivers is not due solely to the want of depth at the bar, but in a great measure to the meeting of the outward current with the waves of the sea, which here form a kind of miniature *roost.* It may therefore be in some rare cases an evil to increase the amount of backwater, as the effect would be to increase the current. The velocities at the entrance of some of our British rivers are given in the following Table :—

Names of Places.	Authorities.	Statute miles per hour.
The Tay near Buddonness .	Admiralty Pilot .	2.88
The Esk near Montrose . .	... ...	7.58
Dee, Aberdeen, between piers .	Min. Civ. Eng. .	8.0 to 9.2
Wear, Sunderland . . .	T. Meik . .	Springs 2.0
Tyne		Neaps 1.25
Mersey, abreast of Helbre and Formby. The same outside of bars	Denham . .	2.75
Mersey, between Seacombe and Prince's Dock . . .	Baines' Liverpool	6.75
Ayr	... ...	.33
Humber	North Sea Pilot .	4.7
Queensferry passage, Firth of Forth	... ...	6 to 7
Clyde at Greenock, last $\frac{1}{4}$ ebb .	Messrs. Stevenson	3.33

When a swell encounters a rapid opposing current, the onward motion of the waves is arrested, and their length is visibly decreased. They get higher and steeper, crest, and at last break, sometimes very partially, and at other times almost as they would on a shelving beach. It is probable that in such disturbed waters several waves may ultimately combine into one very large billow; for one wave may have its onward motion so much checked as to allow the wave behind to overtake it, and the two having thus coalesced, may, as one large wave, acquire a superior velocity, so as to overtake those in front.

The Tides sometimes increase the Surf on the Shore.—It is probably to the velocity of the tide currents, among other causes, that such wonderful effects as that at Whalsey and Wick, already noticed, may be referred. Were such violent action common to all the shores of the German Ocean, instead of being fortunately restricted, as it is, to places where the depth of water is great and the currents strong, some of our eastern seaport towns would, from their low level, be destroyed during the first stormy winter.

Time of the Tide at which the Surf is heaviest.—As a further proof of the great effect of the tides on the waves, it may be stated that the time when most damage is done to sea works which are in tolerably deep water is generally from *one to two hours before and after high-water*, which nearly corresponds to the time when the tide has attained or has begun to attain its full strength. Murdoch Mackenzie, the justly celebrated marine surveyor and hydrographer of the last century, in speaking of the tides of the Orkney Islands, tells us that "the spring tide acquires a considerable degree of strength in less than one hour after its quiescent state begins. Neap tides are hardly sensible in two hours after still water. The stream is most rapid commonly between the third and fourth hours of the tide." * On the 15th February 1853, during a gale from the north-east, a large body of water was thrown upon the lantern of Nosshead Lighthouse, Caithness-shire, being a height of 175 feet above the sea. This occurred *one hour before high water.* On the 23d November 1824, *one hour and a half before high water*, a very alarming wave struck the Eddystone tower, and enveloped the house to a most unusual extent. The mass of water elevated by this wave broke five panes of the lightroom glass. Another remarkable instance occurred at Peterhead Harbour, which projects prominently into the sea on an isthmus, where the tides, at but a short distance seaward of the harbour, run very rapidly. On the 10th January 1849 there was a tremendous sea on the shore, and a crowd of people were down, about *two hours before high water*, helping to secure the whalers and other vessels lying at the quays, when three successive waves, bursting over the harbour, carried away 315 feet of a bulwark founded 9½ feet above high-water springs, and which had stood for many years. One piece of this wall, weighing 13 tons, was moved to the distance of 50 feet. After this violent outbreak of the sea, the waves became more moderate, until about *two hours after high-water*, by which time the large whalers had taken the ground, when other three enormous waves again swept over the harbour, submerging the quays to the depth of 6 or 7 feet, and occasioning the loss of sixteen people, who were washed off the pier. These waves filled the harbour to such a depth as to set all the whalers afloat again, and they continued so for several minutes, until the excess of water had run out through the harbour mouth.

Characteristics of Coasts, the Exposures of which are much

* Orcades, by Murdoch Mackenzie: London, 1750, p. 4.

affected by the Tides.—From what has been adduced in this and another chapter, the following conclusions seem to be warrantable :—

1. The waves are most destructive when they come in at right angles to the shore-line.

2. Their power is increased in proportion as the direction of the main body of the tide approaches to coincidence with the direction of the heaviest swell; and they are probably worst at those headlands on which the tide splits.

3. At parts of the coast where strong tide-currents set off the shore, they reduce the waves by acting as a breakwater.

4. Where a considerable part of the coast retires, there will be less sea during the strength of the tide, even although the waves come in at right angles to the shore, because the tide keeps outside, following the direction of the general *trend* of the coast. But this will probably not hold true of *small* re-entrant hollows of the shore.

5. Although the line of exposure and the tide-current are parallel to the coast, yet if the tide runs in a line *very near* to the shore, as is the case in short narrow channels where the velocity of the current is increased, there may, nevertheless, be an unusually heavy sea.

6. The shores which are most severely tried will probably be those where the line of maximum exposure is at right angles to the line of shore, and where it coincides with the direction of the principal tide-current.

I should not have dwelt at such length on this subject, were it not that I might again refer to some of the facts when treating of the subject of vertical and sloping walls for harbours of refuge, where it is of importance to show that even in the deepest water the waves are not at all times purely oscillatory, but that wherever there is a tide-current the waves will more or less partake of the properties of waves of translation.

Relation between Height of Waves and general Depth of the Sea adjoining.—Another circumstance affecting the exposure of any marine work is the depth of the sea or ocean on the shores of which it is built. The great rolling billows so commonly met with in the Atlantic cannot be generated in the shallower parts of seas like the German Ocean, unless, perhaps, in such peculiar circumstances as have just been adverted to.

Mr. D. Stevenson, in 1838, in his *Engineering of North America** gave it as his opinion that "to the production of considerable undulations, capable of injuring marine works or endangering their stability, three conditions were necessary:—*First,* That the sheet of water acted upon by the wind shall have a considerable area. *Second,* That its configuration shall be such that the wind moving over it in any direction shall act upon its surface extensively, both in the directions of length and breadth. And, *Third,* That the depth of water shall be considerable, and unobstructed by shoals, so as to permit the undulations to develop themselves to a great extent without being checked by the retardations caused by shallow water and an unequal bottom."

In a paper read before the Royal Society of Edinburgh in 1859 I attempted to show, in accordance with those views, that one cause of the peculiarly heavy waves which fall upon the Bound Skerry of Whalsey was not only the great depth of water close to that rocky islet, but the great depth of the German Ocean in those northern latitudes as compared with its southern portions. But as the reduction in the height of the undulations caused by shallow water will be again referred to, I shall only here advise the reader, when judging of any

* Engineering of North America, by D. Stevenson: London, 1838, p. 67.

locality, not to confine his attention to the *local* depth which exists immediately in front of the harbour, but to bear in mind the *general depth* of the sea or ocean on the shores of which his work is to be placed. As examples of the differences in this respect, the following instances are given of different harbours on the same coast :—

Deepest Water at	½ a mile off the Coast.	1 mile off the Coast.	At the Works at low water.
Sunderland	27 feet	52 feet	4 feet.
Tyne	30 ,,	70 ,,	
Dunbar	36 ,,	78 ,,	
Aberdeen	36 ,,	102 ,,	
Peterhead	144 ,,	162 ,,	7 ,,
Wick	102 ,,	135 ,,	34 ,,

Depth in front of a Harbour regulates the Height of the Waves—Largest Waves not always the most destructive.—If the shoal water immediately in front of a harbour extends seawards for a considerable distance, so as to form an extensive flat beach or foreshore, that depth does become the true limit for the maximum wave, whatever may be the general depth of the sea outside. At Arbroath, for example, Mr. Leslie found that the works were in general not so severely tried by the very heaviest class of waves as by others of lesser size. The small depth over the outlying rocks had the effect of *tripping* up the heavier seas, so as to destroy them before they reached the harbour, while it was still sufficient to allow the smaller waves to pass over the shoals and reach the works in an unbroken state. In like manner, at the River Alne, on the Northumberland coast, it is observed that the smaller waves occasion a greater range in the harbour than those larger ones which break outside, and are therefore reduced in passing over the bar.* It thus appears

* North Sea Pilot, Part iii. p. 31: Lond. 1858.

that *the largest waves are not in all places so destructive as smaller ones.* We may also conclude that in cases of severe exposure, where it would not interfere with the passage of ships, the waves might to a certain extent be reduced by dropping very large blocks of stone or concrete at some distance seawards of the works, so as, by forming an artificial shoal, to cause the waves to crest and break outside. In connection with this subject we may state Mr. J. T. Harrison's opinion, that "during violent on-shore gales the water is altogether raised, so that the medium line between high and low water is sometimes raised several feet. The greatest encroachment upon the beach will, on such occasions, take place during neap tides, for it gives the greatest depth of water over the fore-shore at low water."*

It is quite possible, in certain cases, that there may be a very considerable depth at low-water close to the pier, arising from the geological formation, or due to the scouring action of a local current, while the general character of the sea outside may be that of a shallow basin, encumbered with reefs or sandbanks, which render the formation of heavy billows altogether impossible.

Depth of Water in which Waves break.—It is of great importance to be able in all cases to ascertain the maximum possible wave that can exist unbroken in any given depth of water. Mr. Scott Russell, whose observations on what may be called the marine branch of hydrodynamics are of such great value, has stated that "he has never noticed a wave so much as 10 feet high in 10 feet water, nor so much as 20 feet high in 20 feet water, nor 30 feet high in 5 fathoms water; but he has seen waves approach very nearly to those limits." It is presumed that the datum here referred to is the mean

* Min. Civ. Eng. vii. 343.

level of the surface of the sea. As the subject is very important—because the depth of water for some distance in front of a work may be said to be the ruling element which determines the amount of force which it has to resist, whatever be the line of exposure, I shall mention some results that I have obtained on this subject, and which, so far as they go, confirm Mr. Russell's law.

OBSERVATIONS made at the Firth of Forth on Breaking Waves on a sandy beach.

Total Height of Wave.		Depth of Water in Hollow of Wave.	
Ft.	In.	Ft.	In.
2	6	1	2
3	0	1	5
3	0	1	5

It must, however, be borne in mind that these observations, and I conceive also those of Mr. Russell, apply only to those short, steep, and superficial waves, which are due to an existing wind; and not to the ground-swells which are almost constantly to be found in the open ocean, and which may be the result of former gales, or are the telegraph, as they have been called, of those which are yet to come.

Since the first edition of this book was published I had an opportunity, in July 1870, to make observations during a north-easterly ground-swell.* The heights could be measured with very considerable accuracy on the iron piles and open sloping slip or grating at the seaward end of the new iron pier at Scarborough.

* Nature, August 9, 1872.

The following are the results of breaking waves from hollow to crest :—

Heights.
5′ 6″
5 0
5 0
5 6

5′ 3″ = mean height.

The mean depth of water below the trough was 10 ft. 3 in.

Heights of the highest breaking waves from hollow to crest :—

6′ 0″
6 0
8 0
6 0
6 0

6′ 6″ mean height of highest waves. The mean depth of water below the trough was 13 ft. 8½ in., so that in both cases those waves *broke when the depths below their troughs were about twice their own height.* I much regret that I omitted to note the length between the crests. Taking their crests and troughs equi-distant from the mean level of the surface, would give

$$D = 2.5\,h$$

when D = depth of water in feet below the mean level, and h = height of wave in feet from hollow to crest.

On the other hand, some of the large waves in Wick Bay during storms were noticed to break when they came into water of the same depth as their height. The height of those waves above the mean level was about two-thirds of their height, and the hollow below the mean level was about *one-third.*

It is well also to remember that when the bottom shoals suddenly, the waves are more apt to break than when the shoaling is more gradual.

Height of Waves above the Mean Level.—The late Dr. Rankine has shown that the mean water-level is not situated half-way between the crest and trough of the sea. He kindly sent me the formulæ for ascertaining the mean level of the sea from the height and length of the wave. These formulæ are *exact* only for water of considerable depth as compared with the length of a wave. For shallower water they are only approximate.

Let L be the length of a wave.

H the height from trough to crest.

Then diameter of rolling circle $= \dfrac{L}{3.1416}$

Radius of orbit of particle $= \dfrac{H}{2}$

And elevation of *middle level* of wave above still water

$$= \frac{3.1416\ H^2}{4\ L} = .7854\ \frac{H^2}{L}$$

Consequently—

Crest above still water $= \dfrac{H}{2} + .7854\ \dfrac{H^2}{L}$

Trough below still water $= \dfrac{H}{2} - .7854\ \dfrac{H^2}{L}$

CHAPTER VI.

DESIGN OF PROFILE, ETC., OF HARBOURS IN DEEP WATER.

Definition of a Breakwater — Comparison of Vertical and Talus Walls — Oscillatory and Waves of Translation—Wind causes Currents—Forces against Unfinished Masonry—Level of Conservation of Rubble—Best Profile—Comparison of different Works—Oblique Forces vertically and in azimuth—Ratios of Friction of Stones—Russell on Refuge Harbours—Proportions of different Breakwaters — Different Designs — Available Capacity of Harbours of Refuge and Natural Bays.

Definition of a Breakwater.—Harbours of refuge are distinguished from tidal harbours mainly by the superior depth of water which they possess, and the larger area which they inclose. The requisites are—shelter during storms, good holding-ground, and easy access for shipping at any time of tide, and in all states of the weather. A breakwater, though a passive, is yet a real agent, having true work to do. During storms many thousand tons of water are elevated and maintained above the sea level; and these have to be brought down to that level and destroyed within a given space. This is the work which the breakwater has to do. There are two ways in which the work can be performed. One is by means of a plumb wall to alter the direction of the moving water by causing it to ascend vertically, and then to allow it to descend vertically, by which process the waves are reflected and sent back seawards. Another mode is to arrest the undulations by a sloping wall of length sufficient to allow the mass of elevated water to fall down upon the slope. If, however, this slope is not long enough to enable the waves to destroy themselves, they will, though reduced in height, pursue their original direction, and pass over the top of the breakwater. In this case the breakwater does not do its full share of work, and little or no shelter is produced.

Comparison of Vertical and Talus Walls for Breakwaters.—There has been much discussion as to whether piers for harbours of refuge should be *vertical* or *sloping*. Col. Jones, R.E., proceeding on his experience at Kilrush pier, a section of which is given in Plate VII., has especially advocated the superior merits of the vertical wall; and the discussions on his plan at the Institution of Civil Engineers, and the able protest by the late Sir Howard Douglas, will be found, from their interest and importance, to merit a careful perusal.

The principle asserted in favour of the vertical wall is, that oceanic waves in deep water are purely oscillatory, and exert no impact against vertical barriers, which are therefore the most eligible, as they have only to encounter the hydrostatic pressure due to the height of the reflected billows, which are reflected without breaking.

Oscillatory Waves and Waves of Translation.—From the effects of winds and of tide-currents already referred to, and perhaps from other causes, the action of which seems to have been overlooked by the advocates of the upright wall, we have very good reason for believing that any form of barrier, in whatever depth of water it may be placed, must occasionally be subjected to heavy impact. The possibility of waves of translation being generated in the deepest water has been already established in the foregoing chapters, if the reader has been satisfied of the truth of the following assertions:—*First*, That oceanic waves break, partially at least, long before they reach the shore, because (as explained even by the advocates of the purely oscillatory character of oceanic undulations) the depth of water is too small to admit of their propagation;

secondly, That waves in strong tideways break in deep water during calm weather—a phenomenon which is apparent to the eye, and familiar to all sailors ; *thirdly*, and negatively, That to leeward of those races which produce broken water, and which certainly do not reflect the incoming waves, there is comparatively smooth water both at sea and on the adjoining shore until the strength of the tide is exhausted, and the race has disappeared, after which violent action is again fully manifested on the shore.

It may be argued that these are extreme cases, and that such high tidal velocities are seldom met with. This objection has, no doubt, truth in it; but still the tendency is shown, and, though the velocities may be less in other places, there may yet be a current sufficiently strong to destroy the condition of *stagnation* which the oscillatory theory assumes, to say nothing of the impulsive force of the wind. The breaking of waves at sea, and the existence of races, seem to prove beyond question that at least partially breaking waves are possible in the deepest water.

Effect of the Wind in causing Currents.—Wind may also generate currents where there is little tide. It is asserted by Vice-Admiral Zhartmann, in his *Danish Pilot*, that in the Kategat, where the tides have a velocity of 1 to 2 knots, and the common rise is one foot, "the current may sometimes, in boisterous weather, continue to run for three weeks the same way, and even to attain the velocity of 4 knots; and in a furious gale of wind, on the 15th of January 1818, the water rose 5¾ feet above the common water-stand."* He also mentions (p. 260) that in the Great Belt the velocity is increased in south-east storms from 1 or 2 knots to 5 knots in the narrows of Hasselö, and that north and west winds produce

* The Danish Pilot, by Vice-Admiral Zhartmann : London, 1853, p. 122.

similar effects in the Sound ; and he adds, "Nor is it necessary to this result that the last-named winds should blow home : it is enough that a gale should have swept across the North Sea in that direction for several successive days" (p. 134). If these statements be correct, they certainly prove that the water is driven with force by the wind.

Damage done by Waves in Deep Water.—In 1858 the "Hebe" of Wisbeach lost her bulwarks in her passage from Sunderland to Leith, and on examining the crew immediately after her arrival, the sailors stated that the *weather* bulwarks were the first that were stove in. It will surely be admitted, that in this case, besides the impact due to the velocity of the vessel, the waves must have exerted an impulsive force, and that not due to the action of the wave *after* but before it had broken on the deck. Is it not also probable that even purely oscillatory undulations, after being reflected by a vertical wall, may combine with others of the same kind so as then to become waves of translation, possessing all the elements which endanger the stability of a sea work?

Dr. Scoresby, in narrating his experience of the storm which he encountered on his voyage to Melbourne in the "Royal Charter" in 1856, gives a graphic account of the breaking of oceanic waves. He says,* "No wave could keep pace with the legitimate demands in hydrodynamic law of the wind's terrible vehemence. Waves of 40 feet in height, which satisfy the greatest demands perhaps of any of our North Sea or high northern Atlantic storms, bore no adequate relation to the impetuosity of this hurricane tempest. A sort of surface impetus seemed to be given, forcing the crests of the loftier waves with a velocity so much beyond the motion of the regular undulations, as not only to cast almost every peak

* Life of William Scoresby, by Dr. Scoresby Jackson : Lond. 1861, p. 374.

and summit into the form of a breaker, but in some cases to give such a degree of magnitude and breadth to the breaking summit—as one mass of white water labouring forward after another, and retarded by the diminished velocity of that before it—that the main surface *behind* some of the mightiest waves would present but one unsubdued and wide-spread breast of foam—a phenomenon I had never seen but in waves breaking over an insulated shelving rock!"

Waves of Translation formed by the Unfinished Masonry of Deep-water Works.—But whatever value may be attached to the facts which we have adduced as to the certainty of deep-water waves exerting, in some circumstances, more than the hydrostatic pressure due to their height, there is another consideration, the cogency and relevancy of which can hardly admit of any question. It is a well-known and generally admitted fact, that damage to marine works occurs, in the great majority of instances during their progress, and not after their completion. Every contractor knows the advantage of having a sea work finished before the storms of winter commence, and it is usual to *close-in* the works with a temporary wall when the autumn draws to an end. Any one may see the obvious propriety of such a course, for it is plain that the stones which are last set have only their own weight to keep them stationary, while in a pier that is finished, the materials are not only bound at both ends, but are kept in their places by the weight of the superjacent masonry. When a storm comes suddenly on during the progress of the works, it at once overturns and removes the outermost stones, and as these expose others, they too, in their turn, speedily disappear, while the mischief is aggravated by the washing out of the still more easily moved backing and hearting. It is manifest, therefore, that this piecemeal destruction will not be confined to the open end of the pier, but will encroach upon the finished work.

Now, it is equally well known, that in order to preserve the bond in a wall of masonry while in progress, the unfinished end must always present a stepped or serrated outline. It follows, therefore, that even a vertical wall must, during its formation, present constantly to the action of the waves a sloping or at least a stepped face like a talus wall, but which, unfortunately for its stability, possesses none of the advantages of a finished sloping breakwater. In short, during the most critical period of the history of every vertical wall, the face-work and hearting are exposed at the outer end to the force of *breaking* waves, which not only act upon the materials, but act upon them at the very time when they are in the most defenceless state.

Dynamometrical Values of Force against Unfinished Masonry.

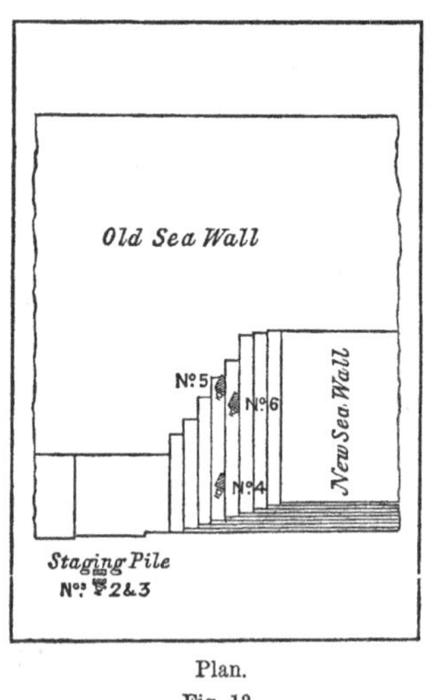

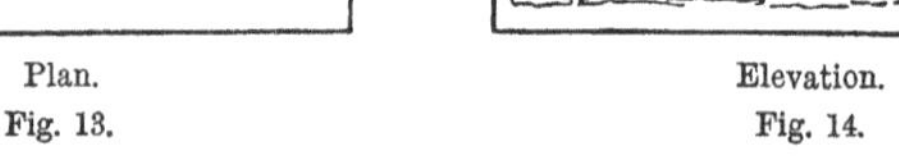

Plan.
Fig. 13.

Elevation.
Fig. 14.

—In October 1858, in order to bring these views to the test

of experiment, I fixed small Dynamometers on an unfinished wall at the harbour of Dunbar. The works consisted of the formation of a new sea-wall in front of an old one which had become ruinous. Seaward of the new wall the contractor erected piles for supporting a travelling crane. Two Dynamometers, marked Nos. 2 and 3 in Figs. 13, 14, were fixed to the piles, while other three, Nos. 4, 5, 6, were fixed on the unfinished wall. Mr. Russell's law, that waves do not break or become entirely waves of translation until they reach water so shoal as to be no deeper than their own height, appears to hold good for the smaller class of waves; and, therefore, in depths of from 7 to 11 feet 5 inches, being the depths at high water in front of the works, while the trials were made, waves of from 16 inches to 3 feet 9 inches would reach the wall without becoming waves of translation; while waves of from 7 to 10 feet, being those which formed the subject of trial in the first part of the table of experiments given below, would in that depth become breaking waves.

On consulting the Table of Observations, it will be seen that the mean force registered on the unfinished courses of the new sea wall, including those imperfect results in which the spring was driven home, was with waves of translation 2.01 times greater than was registered on the pile. This is much what might have been expected, for the water in motion would escape more readily past the pile, which presented a very small surface to the moving water, and allowed it to pass freely on both sides. But if we look to the second part of the table we find that the force with the oscillatory or non-breaking waves is 22.45 times greater on the unfinished courses of the masonry than on the pile outside, showing clearly that the broken surface of the unfinished vertical wall had changed their character from oscillatory waves to waves of translation.

OBSERVATIONS made in 1858 at Dunbar Harbour on the Effects of Waves of Translation and of Oscillation against Dynamometers fixed on the unfinished wall, and on isolated piles in front of the wall. The effects are those produced during each tide, the results having been read off at each low water.

Date of Observation.	Depth at high water. in feet.	Waves of Translation.										Height of waves in feet.
		No. 2. On Pile 7 feet above bottom.		No. 3. On Pile 9 feet 7 inches above bottom.		No. 4. On Sea-wall 8 feet above bottom.		No. 5. On Sea-wall 9 feet 6 inches above bottom.		No. 6. On Sea-wall 11 feet 2 inches above bottom.		
		Force in cwts.	Position of Dynamometer.	Force in cwts.	Position of Dynamometer.	Force in cwts.	Position of Dynamometer.	Force in cwts.	Position of Dynamometer.	Force in cwts.	Position of Dynamometer.	
1858, Nov. 5	$10\frac{1}{4}$	5.09	$3\frac{1}{4}$	12.48	$\frac{1}{2}$	6.53	$2\frac{1}{4}$	† 45.04	$\frac{3}{4}$	11.81	1′ dry	10
6	10	6.89	3	9.70	$\frac{1}{4}$	13.06	2	15.76	$\frac{1}{2}$	7.09	1′ 3″ dry	7
24	10	8.42	3	† 13.61	$\frac{1}{4}$	† 17.61	2	† 23.55	$\frac{1}{2}$	19.27	1′ 3″ dry	10
25	9	.94	2	4.76	$\frac{3}{4}$ dry	† 17.61	1	† 23.55	$\frac{1}{2}$ dry	21.67	2′ 3″ dry	7
27	8	† 9.36	1	† 13.61	1′ 9″ dry	3.52	0	† 23.55	$1\frac{1}{2}$ dry	6.02	3′ 3″ dry	7
		30.70		54.16		58.33		131.45		65.86		
	mean	6.14		10.83		11.67		26.29		17		

OSCILLATORY WAVES.

Nov. 18	$9\frac{3}{4}$	0.0	$2\frac{3}{4}$	0.0	0	0.0	$1\frac{3}{4}$	40.53	$\frac{1}{4}$	°——	1 6 dry	$3\frac{3}{4}$
19	$10\frac{1}{4}$	0.0	$3\frac{1}{4}$	0.0	$\frac{1}{2}$ dry	6.53	$2\frac{1}{4}$	22.52	$\frac{3}{4}$	°——	1 dry	$1\frac{3}{4}$
Dec. 3	$8\frac{1}{2}$	1.40	$1\frac{1}{2}$	°——	$1\frac{1}{4}$ dry	0.0	$\frac{1}{2}$	14.13	1 dry	°——	$2\frac{3}{4}$ dry	2
4	$8\frac{1}{2}$	0.0	$1\frac{1}{2}$	6.80	$1\frac{1}{4}$ dry	14.08	$\frac{1}{2}$	21.19	1 dry	°——	$2\frac{3}{4}$ dry	2
6	$10\frac{1}{4}$	0.0	$3\frac{1}{4}$	0.0	$\frac{1}{2}$ dry	4.40	$2\frac{1}{4}$	8.24	$\frac{3}{4}$ dry	°——	1 dry	$1\frac{1}{2}$
7	10	0.0	3	0.0	$\frac{1}{4}$ dry	7.40	2	14.13	$\frac{1}{2}$ dry	°——	$1\frac{1}{4}$ dry	$1\frac{1}{2}$
8	$9\frac{3}{4}$	0.0	$2\frac{3}{4}$	0.0	0 dry	7.40	$1\frac{3}{4}$	18.84	$\frac{1}{4}$ dry	°——	$1\frac{1}{2}$ dry	$1\frac{1}{2}$
9	9	0.0	2	0.0	$\frac{3}{4}$ dry	0.0	1 dry	21.19	$\frac{1}{2}$ dry	°——	$2\frac{1}{4}$ dry	1
11	$7\frac{3}{4}$	0.0	$\frac{3}{4}$	°——	2 dry	3.52	$\frac{1}{4}$ dry	°——	$1\frac{3}{4}$ dry	°——	$3\frac{1}{2}$ dry	2
15	6	0.0	$\frac{1}{2}$ dry	°——	$3\frac{1}{4}$ dry	2.64	$1\frac{1}{2}$ dry	°——	3 dry	°——	$4\frac{3}{4}$ dry	$2\frac{1}{2}$
		1.40		6.80		45.97		160.77				
	mean	.14		.97		4.59		20.096				

Note.—The sign † shows that the spring has been driven home, and that the result indicated may not be the maximum.

The sign ° shows that these observations are inapplicable from being above the reach of the waves, and must not be included in calculating the means.

It must be remarked, however, that the dynamometer No. 5 was placed purposely in the angle formed by the junction of the new and old walls, where, as might have been expected, the force was concentrated. This will be perceived on comparing its indications with the others. Although in some parts of every unfinished wall re-entrant angles such as this must exist, where similar concentration of force will take place; yet, even although we exclude its indications from the result, it still appears that the ratio for waves of translation is only 1.46 times greater than that on the log, while with oscillatory waves it is 8.27 times greater.

These experiments prove therefore that *oscillatory waves become waves of translation when they reach the unfinished part of a vertical sea wall, and that they then exert a force nearly* 6 *times greater than if they had remained waves of oscillation.* It farther appears that purely oscillatory waves do not exert much more than their hydrostatic pressure, under circumstances similar to those affecting the Dunbar experiments; but had there been a storm of wind, those waves would no doubt have ceased to be purely oscillatory, even although the water had been very deep.

Best form of Breakwater. Level of Conservation of Rubble.— If we further keep in view that any settlement of the foundation is far more perilous to a vertical than to a sloping wall, there seems good ground for believing that the ordinary method of forming the low water parts of deep harbours of large masses of rubble stone or of concrete blocks, is, in most circumstances, the best and cheapest kind of construction when a vertical wall is to be adopted. Loose rubble or blocks of concrete, after being acted upon by the waves, are less liable to sink, or to be underwashed, than when a vertical wall is founded upon a soft bottom. Loose concrete blocks above low water form an excellent protection to the upright wall. Two precautions should, however, be kept in view—*First*, the wall should be founded at a sufficiently low level to prevent underwashing. A depth of from 12 to 15 feet under low water was pointed out by the late Mr. J. M. Rendel as the level below which the waves did little or no damage to *pierres perdues.* Sir John Rennie indeed considered that there was little or no effect at a fathom and a half.* No works hitherto executed had, so far as I know, been founded at a lower level than 12 feet, but at Pulteneytown, where the rubble is more exposed than at any other harbour, Messrs. Stevenson, as already stated, considered it advisable to found the wall 18 feet below low water, and the rubble has actually been removed down to 15 feet below low water. *Secondly,* in all cases where the structure is to act simply as a breakwater, and not as a pier, there should be no parapet, the want of which relieves the foundation, as was observed by Mr. D. Stevenson at a harbour work where a breach had been made. At one part, where the wall remained entire, and the sea was opposed by the parapet, the sea fell heavily on the foundations; but at the breach, where the parapet was wanting, the waves played gently over the work, without any perceptible reaction against the foundation courses. Mr. Murray also approves of there being no parapet, and proposes that the roadway of breakwaters should be 10 feet above high water. When pitched slopes are adopted, great benefit will be found to accrue from leaving a wide fore-shore at the bottom or toe of the slope. Much, however, depends on local peculiarities in selecting the best design for any work; and the nature of the bottom is in all cases important. Where the bottom is soft, a high vertical wall should not be attempted.

Comparisons of different Works.—In making these re-

* Min. Civ. Eng. vol. vi. p. 122.

marks, I must not be understood as condemning the adoption of vertical walls without a rubble base, in cases where the foundation is good. All that is asserted is the opinion that partial waves of translation do exist in deep water, and therefore that harbours of refuge will prove failures unless they are built of sufficient strength to resist the impact of such waves. The Cherbourg breakwater (Plate VII.) has been often referred to as a successful instance of the application of a vertical wall, and has been contrasted with the Plymouth breakwater, which has a long slope. But this appeal is quite fallacious, as the Cherbourg breakwater is of a composite character, consisting of a mass of rubble sloping at the rate of 7 horizontal to 1 perpendicular, surmounted by a plumb wall ; so that, whatever merit may be supposed to belong to the vertical profile, is entirely nullified by the long slope in front, on which the waves break before they reach the vertical barrier. Moreover, the heaviest waves at Cherbourg come from the N.W., and do not assail the breakwater at right angles to its direction, but come more nearly *end on* to the work, so as to a great extent to run along the outer wall. The N.W. waves are propagated from the Atlantic, while those which prove most trying to the work come from the N., in which direction the extent of exposure is only about 60 miles. These facts were obtained during a visit to Cherbourg, undertaken for the special purpose of ascertaining the physical characteristics of the place. Any attempt to establish a parallelism between Plymouth, which faces the Atlantic directly, and Cherbourg, which is comparatively land-locked, cannot, therefore, stand the test of a careful inquiry.

Other comparisons may be referred to which have been advanced on equally untenable grounds. Thus, the old pier of Dunleary, which is vertical, and has stood well, has been compared with the talus walls of Kingstown Harbour, which now protect Dunleary, and which have often received much damage. The all-important element of the *depth of water* in front has in this comparison been entirely overlooked; for at Kingstown there is a depth of 27 feet, while Dunleary is all but dry at low water.

Oblique Force in the Vertical Plane.—An important advantage of the sloping wall is the small resistance which it offers to the impinging wave, but it should also be borne in mind that the weight resting on the face-stones in a talus wall is decreased in proportion to the sine of the angle of the slope. If we suppose the waves which assail a sloping wall to act in the horizontal plane, the component of their impulsive force at right angles to the surface of the talus will be proportional to the *sine* of the angle of inclination to the plane, while the effective force estimated in the horizontal plane will be proportional to the *square of the sine* of the angle of inclination. But if we assume the motion of the impinging particles to be horizontal, the number of them which will be intercepted by the sloping surface will be also reduced in the ratio of the sine of the angle of inclination, or of the inclination of the wall to the vertical. Hence the tendency of the waves to produce horizontal displacement, on the assumption that the direction of the impinging particles is horizontal, will be proportional to the *cube of the sine of the angle of elevation of the wall.*

Formula for Oblique Force vertically and in Azimuth.—If it further happens that, owing to the relative direction of the pier and of the waves, there is an oblique action in azimuth as well as in the vertical plane, there will be another similar reduction in the ratio of the squares or cubes of the angle of incidence, according as the component of the force is reckoned at right angles to the surface of the pier, or in the direction of the waves.

Let $f =$ force of the wave on unit of surface of wall for perpendicular incidence ;

$f' =$ force on unit of surface at vertical incidence φ, and azimuthal incidence ψ ;

then $$f' \propto f (\sin \varphi \sin \psi)^3.$$

Friction of Stones on each other.—The above expression assigns perhaps too great a reduction to the oblique action, because the motion of many of the particles is not horizontal, and the upward force acting over the area of the lying beds of the stones is perhaps more to be dreaded when the works are in progress than the horizontal force against the outer vertical faces. Yet it must be observed that the experience at Wick Breakwater, as afterwards referred to, points, on the other hand, to a different conclusion.

Ratios of Friction for different kinds of Masonry.—More extended experiments than have yet been made are desirable to determine the *constant* for the friction of rough blocks over each other. Mr. George Rennie found that $\frac{8}{10}$ths of the weight were required to drag a block of stone over the roughly-dressed floor of a quarry, and that the voussoirs of the London Bridge began to glide over each other at a slope of from 33° to 34°.* I made a few experiments on the friction of small, polished, ashlar blocks of freestone, and the mean gave $\frac{5}{10}$ths of the whole weight moved, as the coefficient of friction of such stones over a similarly polished freestone block.

The power required to *extract* a polished freestone block out of its place in a column consisting of different numbers

* Phil. Trans. 1829, p. 168.

of blocks was also tried. The coefficient of friction in extracting any of the blocks from the column was 1.083, or about *twice* the amount for moving the whole mass (including the stone extracted).

As the blocks in a sea-work are very often submerged, it appeared desirable to ascertain the friction in water as well as in air, and the following table shows the results, as also the coefficients for stones in different styles of dressing.

Kinds of Dressing and of Materials.	Angles at which blocks began to move.		Coefficients of Friction for Water.	Coefficients of Friction for Air.
	In Water.	In Air.		
Polished freestone on polished freestone	28°	28°	.53	.53
Longitudinal axing on cross broaching*	...	37½°		.77
Cross axing on cross broaching .	...	38¼°		.79
Closely axed greenstone on scabbled freestone	39°	35°	.81	.70
Longitudinal broaching on pick-dressing	...	39¼°		.82
Closely axed greenstone on scabbled freestone	...	40°		.84
Cross close axing on pick-dressing .	41⅓°	40⅛°	.88	.84
Longitudinal broaching on longitudinal broaching	41°	41⅓°	.87	.88
Closely pick-dressed freestone on closely pick-dressed freestone . .	51°	43°	1.23	.94

The blocks which were used in these experiments were of too small sizes to give anything more than a rough comparative valuation for air and water, but some stones of about a ton in weight were dressed for me in different styles of workmanship, at the sight of Mr. Robert Kinnear, Inspector of Works, when the following results were obtained :—

* *Broaching* is a style of work peculiar to Scotland, and consists of a number of narrow parallel ridges running close to each other. They are made with a sharp-pointed tool, and extend over the whole of the face-work between the *drafts*.

Kind of Dressing.	Angles at which Stones began to move in Air.	Coefficients of Friction in Air.
Droving on droving, waves parallel to line of slip	33° 8′	.65
Broaching, on broaching with strips or ridges running parallel to the line of slip . . .	34° 3′	.68
Droving on droving, waves cross to line of slip .	35° 57′	.73
Broaching on broaching, with strips or ridges lying crossways to the direction of slip . .	36° 6′	.73
Rough pick-dressing on rough pick-dressing .	38° 11′	.79

Evil of Fine Workmanship in Harbour Masonry.—These tables are given not as determining the different coefficients exactly, but as establishing the fact that the friction in water is much the same as in air; and as showing the impropriety of expending labour in dressing finely the materials for harbour masonry. By polishing the beds we at once reduce the power of resistance by about *one-fourth;* or, in other words, roughly dressed materials of *three-fourths* of the weight of polished materials will be equally safe, while they are also more economical, both as to the cost of the stone and of the workmanship in dressing.

Mr. Russell's Remarks on Harbours of Refuge.—I have already referred to the importance of Mr. Scott Russell's experimental inquiries, and I cannot do better than conclude the remarks on this subject by a quotation from his observations during a recent discussion at the Institution of Civil Engineers, the value of which will, no doubt, be regarded by the reader a sufficient excuse for its length.

"In sea-works there were practically two classes of waves to deal with, of such different, if not opposite, natures, that what was beneficial in the one case was often useless in the other. * * * * In deep water, there were not only the oscillating surface-waves to be encountered, but also those which he had termed waves of translation, forming what were called rollers at the Cape of Good Hope, and when on a smaller scale, known as ground-swell. These were a much more troublesome class of waves; it was mainly with them that the engineer had to deal, in places open to the Atlantic; and after a storm of some duration at sea, they became the deadliest enemies, in the cases of deep water, against which breakwaters for harbours of refuge had to contend. These rollers, or ground-swell, did not merely oscillate up and down, but backwards and forwards; and they could not be eluded, or turned back, by giving to the wall a particular curve, suited to the form of a cycloidal oscillation.

"These great waves of translation constituted a vast mass of solid water, moving in one direction with great velocity, and this action was nearly as powerful at a great depth as at the surface. They resembled the tidal bore of the Hooghly, of the Severn, or of the Dee; they formed a high and deep wall of water, of great weight, moving horizontally with great force, and causing all floating bodies they met with to travel with them with great velocity in the same direction. As he had before mentioned, they could not be eluded, or diverted, they must be stopped; and therein consisted the difficulty. The only certain way of effecting it, was to oppose to these waves a mass of matter so much heavier than themselves, that they could not move it. By so doing, the waves were compelled either to roll over the obstacle, in which case they would create a new wave inside; or they must be made to break on themselves backwards, which required enormous power; or they must be completely reflected, which, perhaps, required the greatest force of all.

"To reflect or send back the roller was the most effectual

plan. For this purpose, nothing more was necessary than a deep perpendicular face of perfect masonry, and so long as it stood firm, it was faultless, and the water inside was smooth as in a mill-pond ; for the reflection really converted the whole effect of the roller on itself into a simple pressure of water. When such a wave was reflected on a perpendicular wall, it merely produced a hydraulic pressure, equal to that due to little more than double its own height. A roller, 20 feet in height, would produce a pressure of about a ton per foot, and it would be reflected by a vertical wall of moderate dimensions. He had, therefore, no hesitation in saying, that, cost apart, a wall of vertical masonry was the best while it lasted. The primary cost of erecting a vertical wall of perfect materials was, in most cases, so great as to put it out of the question, setting aside the important point of durability, which was also cost in a serious form. A vertical wall of masonry had, however, this great disadvantage; if the sea found out a weak place, it would enlarge it much more rapidly than in an inclined wall. He had seen a stone of small size show symptoms of crumbling at the beginning of winter, and in one week the little hole of single defective stone was converted into a circular breach, nearly 30 feet in diameter. An inclined wall, he had ascertained, would reflect a roller, or deep wave, nearly as well as a vertical wall, down to an inclination of 45°. This observation was of value, because at 45°, large blocks, judiciously placed, would not move out of position, even although a considerable breach was made in the wall near them. He had carefully watched the action of the sea when approaching this slope, and his opinion was favourable to the trial of a wall of heavy rubble blocks fairly laid, so as to form a tolerably even face, or slope of 45°, when the nature of the materials and the local circumstances would permit.

"When any approximative attempt at reflection of the wave, by walls not too remote from the vertical, had to be abandoned, recourse must generally be had to breaking the wave. This was a formidable thing to attempt on a deep-sea roller; and it could only be done by opposing mass to mass. For this purpose he had ventured to recommend that the forefoot of the beakwater should be rounded off, and that the shape of the breakwater should be convex. The intention of this was, to cause the front wave to begin breaking at the earliest moment, to make the breaking last as long as possible, and thus to render the diminution of its momentum as complete as was practicable. In breaking a wave, it was very important that it should be made to break on the water, and not on the stones; and the convexities given to the foreshore accomplished this object, by causing the waves to begin breaking as far out as possible in deep water. The form for breaking a deep roller should be entirely distinct from that used for meeting a superficial oscillating wave; the one should be concave, the other convex. The Digue, or breakwater of Cherbourg,* showed an approximation to this form. A breakwater, to succeed in breaking the long waves of translation, rolling in from deep water after a storm, should have a long convex slope.

"The practical construction of breakwaters was, however, scarcely confined either to resisting deep-sea rollers, or waves of translation, or to procuring the means of stopping the progress of common sea waves, or mere surface oscillation; it usually combined both of these objects. It might, therefore, be considered as the general problem of a breakwater—first, to stop out the great wave of translation, and secondly, to still the oscillating surface-wave. A vertical wall effected

* *Vide* Plate VII.

both these conditions, and so also did a convex sea-slope with a vertical pier. But the best plan of all was, first to carry up a convex sea-slope of rubble to near the surface of the water, and thus break the force of the heavy ground-swell, that would sweep ships from their anchors, and lay them high and dry on a lee shore. This would allow the wave to break to pieces on itself, and to expend its force in raising so much water as represented its momentum to a height above its former place sufficient to exhaust its power; in other words, to expend its power on water. Secondly, to carry up from the top of the rubble slope a wall slightly inclined, to reflect the waves of oscillation near the surface, which could do no harm if quietly resisted and sent back. This plan seemed to be that which had been found to answer best in practice. He further considered, that making a step backwards in the upper work of a breakwater was of great value in preventing the tops of the waves from going over. He had carefully watched the effect at Marseilles, and his observation had confirmed the opinion he expressed in favour of it in 1847.

"These views of the nature of the forces to be resisted in breakwaters, and the methods of dealing with them, showed the inefficiency of what were called floating breakwaters Large floating masses would certainly intercept oscillating waves of a small depth, and in moderate weather they would often still the water. The lee side of the "Great Eastern," when lying at Holyhead, afforded excellent anchorage for small vessels, in light breezes. But when the great roller, the one great wave of translation, came, the anchors snapped at once, showing the danger which would have been incurred had she been moored broadside to the roller, instead of offering to it the small resistance of her fine bow. No known force could effectively secure a large floating breakwater broadside-on to a heavy ground-swell. It would move horizontally with the wave of translation, which would propagate itself along the bottom, just the same as if the breakwater was not there."

Examples of Rubble Breakwaters.—According to Sir John Rennie (*Account of Plymouth Breakwater*), rubble breakwaters with slopes formed at the angle of repose, were adopted by the Greeks in the moles of Tyre and Carthage, and by the Romans at Athens and Halicarnassus. The same design was also followed at Venice, Genoa, Rochelle, Barcelona, and other places. In this kingdom the first example on a large scale which we find is at Howth. Kingstown (Plate VII.), Holyhead, and the noble breakwater at Plymouth (Plate VIII.), were afterwards carried out on the same principle, chiefly under the directions of the late Mr. Rennie. The great national harbours of refuge at Holyhead and Portland (Plate VIII.), formerly under the late Mr. Rendel, and latterly under Sir John Hawkshaw and Sir John Coode, are on a similar principle ; while those of Messrs. Walker and Burgess, at Dover and Alderney, are either nearly vertical or composite (*vide* Plate VIII.)

Proportions of Breakwaters.—The following table of the principal proportions of some of the most remarkable breakwaters may be found useful as a guide in designing works of a similar kind :—

Table of Proportions of different Breakwaters.

Name.	Kind of Work.	General Slope of Outer Face.				Inner Slopes.			Level of top of loose rubble below low water.	Level of foundations of wall below low water.
		From Bottom to near Low Water.	Near Low Water.	Up to High Water.	Above High Water.	Above High Water.	Below Low Water.	Top above High Water.		
					Sloping Breakwaters.					
Plymouth.	Pitched slopes above H. W., loose rubble below.	$1\frac{2}{3}$ to 1	4 to 1	5 to 1	5 to 1	2 to 1	2 to 1	3 feet.	0	0
Portrush.	Slopes of loose rubble.	$1\frac{3}{4}$ to 1	6 to 1	3 to 1	$1\frac{1}{3}$ to 1	$1\frac{1}{2}$ to 1	1 to 1	16 feet.	0	
Kingstown.	Pitched slopes of rubble.	$1\frac{3}{4}$ to 1	5 to 1	5 to 1	5 to 1	$\frac{1}{3}$ to 1	$\frac{1}{3}$ to 1	15 feet.	0	0
Holyhead.	Slopes of loose rubble.	1 to 1	$5\frac{1}{2}$ to 6 to 1	$5\frac{1}{2}$ to 6 to 1						0
					Composite Breakwaters.					
Portland.	Slopes of loose rubble, with plumb wall above high water.	$1\frac{1}{4}$ to 1	5 to 1	5 to 1	5 to 1, & plumb wall.	Plumb wall.	1 to 1	25 feet.	0	
Cherbourg.	Slopes of loose rubble, with plumb wall above.	2 to 1	7 to 1	7 to 1	to 1	$7\frac{1}{2}$ to 1, & $\frac{1}{8}$ to 1	1 to 1	$12\frac{1}{2}$ feet.	0	0
Alderney.	Slopes of loose rubble, with plumb wall above.	2 to 1	5 to 1	Wall $\frac{1}{2}$ to 1	$\frac{1}{2}$ to 1, & plumb	Wall $\frac{1}{3}$ to 1	$1\frac{1}{4}$ to 1	25 feet.	0	12
Cette.	Do.	$1\frac{1}{2}$ to 1, & 6 to 1	$3\frac{1}{2}$ to 1	Plumb wall.	Plumb wall.	..	1 to 1	$12\frac{1}{2}$ feet.		
Pulteneytown.	Do.	From bottom to 15 ft. below low water. 1 to 1, & 7 to 1	..	Do.	Do.	..	..	6 feet, and 21 feet.	15	18
					Vertical Breakwaters.					
Dover.	Solid Masonry.	$\frac{1}{4}$ to 1	$\frac{1}{4}$ to 1	$1\frac{1}{4}$ to 1	$\frac{1}{4}$ to 1, & cavetto.	$\frac{1}{4}$ to 1	$\frac{1}{4}$ to 1	23 feet.		45
Aberdeen.	Concrete.	$\frac{1}{8}$ to 1	$\frac{1}{8}$ to 1	$\frac{1}{8}$ to 1	$\frac{1}{8}$ to 1	$\frac{1}{8}$ to 1	$\frac{1}{8}$ tol	11 feet.		20

Relative economic Values of Different designs for Breakwaters.—I have collected the following costs of different breakwaters from the Minutes of the Institution of Civil Engineers, and other sources.

Name of Breakwater.	Depth of water in fathoms.	Cost per lin. foot.	Cost per lin. yard.	Remarks.
Joliette, Marseilles	5 to 6	£72	£215	No rubble, all large beton blocks.
Algiers	6 to 9	122	366	
Holyhead . . .	5 to 7	About £160	480	
Marseilles (new) .	5 to 6.6	109	328	Convict labour.
Portland . . .	8 to 10	116 to 120	348 to 360	
Alderney . . .	3.3	170	510	
Dover	6.6 to 8.3	360	1080	
Plymouth . . .	6.6 to 7.5	200	600	

Although some of these prices have given rise to lengthened discussions as to the comparative economic advantages of the various designs, I fear that the results have not been of much value, on account of the different degrees of exposure and of depth of water at the various places. The economic values may perhaps be arrived at in a more satisfactory manner, although still but only very approximately, thus:— When x = the price per foot of depth, p = the price per lineal foot, and d = the depth in feet at high water.

$$x = \frac{p}{d}$$

The results calculated in this manner are arranged in order of their costs in the following Table:—

Name of Harbour.	Depth in feet at high water.	$\frac{p}{d}$ in £ and dec.
Portland	62	£1.90
Joliette	35	2.06
Algiers	42	3.00
Marseilles (new) . .	35	3.10
Plymouth	58	3.45
Holyhead	36	4.40
Alderney	37	4.60
Dover	38	9.47

From this view of the subject the plumb pier of Dover appears to be by far the most costly.

Different Projects for constructing Harbours of Refuge.—After the general opinions that have already been expressed, and in anticipation of what follows, regarding the action of the waves, and the choice and relative durability of different materials, it is not necessary to describe the projects for constructing harbour works which have from time to time been brought forward by amateur engineers. So far as my experience has gone, I incline a good deal to the opinion, that until chemistry discloses to us some new process for uniting together or for protecting our materials, we need not look for any *royal road* to harbour-building, though I should be far from discountenancing any attempts at improvement, from whatever quarter they might come; and here I would remark, what will be afterwards referred to, that the hydraulic properties of Portland cement go far to meet the difficulties arising from want of continuity in marine masonry, at least above the level of about half-tide.

The same remarks as to novel designs do not of course apply to others made by those who are professionally acquainted with the subject; but I have no wish to sit in judgment on the merits of any *individual* design, whether made by engineers or others, as such a course might lead me to express unfavourable opinions on some points. I will therefore only give a very brief description of some of these, leaving the reader to form his own judgment as to their relative merits.

Vertical Wave-Screens.—Isolated piles of timber or iron, placed at certain distances apart, have been proposed as breakwaters by Captain Vetch, R.E., and Captain Calver, R.N.

Captain Calver proposes to form a breakwater, or, as he terms it, a wave-screen, consisting of a row of vertical piles driven into the bottom at certain distances apart. They are to be united horizontally by iron runners, and supported laterally by sloping struts having large iron shoes attached at their lower ends for entering and abutting against the bottom, while their upper ends are proposed to be connected with the main upright piles by swivel joints, to admit of their being placed at any angle. The shorter of the struts is to abut upon the main upright a little above the level of low water, while the longer one is to abut upon the main pile at a corresponding height above high water. A short chain extends to the longer strut from near the place where the lower strut joins the main upright. At the top the main piles are connected together by a narrow continuous gangway with hand-rail outside. The piles are proposed to consist either of timber or of wrought iron. The screen is intended for a low-water depth of 36 feet, a tidal rise of 15 feet, and a maximum wave of 15 feet. The gangway is to be elevated about 14 feet above the undisturbed high-water level. The shoes of the struts are to be of a peculiar shape, so as to present a vertical flat surface to the soil in which they are buried. These struts are not to be driven into the soil, but are to be "buried beneath the surface by the tidal current." Captain Calver anticipates that the strength of this framing will be greatly more than sufficient to withstand "any force that could possibly be brought against it." For a full description of the structure and of the merits claimed for it, the reader is referred to Captain Calver's "Wave-Screen," published at London in 1858, which contains much interesting information relating to the subject of harbours.

Horizontal Wave-Screens.—Mr. Brunlees recommends a breakwater and pier of cast and wrought iron. The piles are intended to be placed zig-zag, at an angle of 90°, with the view of increasing the strength of resistance. The piles are to be sunk, where the soil is sandy, by the hydraulic process, which was successfully used by him for sinking the Morecambe Bay Viaducts. The spaces between each pile are to have horizontal louvre boards attached, and, as the spaces are short, the louvre boards are intended to be of cast-iron. The space between low water and the bottom is left comparatively open, with the view of avoiding interference with the run of the tides.

Mr. M. Scott has suggested a combination of horizontal wave-screens with a rubble base; the construction and advantages which he claims for it are thus described by him:*—

"In the case of deep water, and where stone is to be readily obtained, a bank of rubble might be deposited, rising to within, say 15 feet, or more, of low-water mark, the height of the bank varying with the circumstances of the locality. Upon this bank it is proposed to build a face wall, up to low-water mark, and behind this wall, long counterforts, the upper surface of which would rise from low water, at an inclination of about 2 to 1, and extend back for a distance dependent upon the amount of slope rendered necessary by the magnitude of the waves. These counterforts would be placed at sufficient intervals, say of 20 feet, so as to be conveniently spanned by iron girders, and the whole of the sloping surface would be converted into a sort of gridiron, by girders laid from pier to pier, the upper flanges being about 1 foot wide, and the girders laid at intervals of about 18 inches.

"Supposing such a breakwater to be erected and exposed to a heavy sea, if the waves are not breaking, the water would be projected up the slope, and would drop through the spaces between the girders; and if the waves are breaking, they will

* Minutes of Institute of Civil Engineers, 1860, p. 649.

rush up the slope as a confused mass of water, dropping through on their passage. But, although it is anticipated that the great bulk of the water would pass through the grating and not return to the foot of the slope, the operation would be gradual and be diffused over a considerable surface, so that a wave would not be reproduced inside. The only effect would be a stream of water into the harbour, and in this particular, the proposed form differs in principle from all vertical screens, or gratings, which, by permitting the waves to pass through, at the same instant of time, have not the effect of destroying the undulation. The breakwater should, if possible, be placed at an angle with the direction of the greatest sea, so that a wave should not only run up the slope, but along it, diffusing itself in this manner over a much larger area."

Timber Breakwaters in Deep Water.—Mr. Abernethy and Mr. M. Scott have proposed to apply to harbours of refuge in deep water the timber-box principle, which has been long in use for tidal harbours, to which reference will afterwards be made, and of which drawings will be found in Plates IX. and X. Mr. Scott suggests that the structure may be formed by simply resting frames of timber upon the bottom, without having any piles driven into the ground, a plan which has been carried into execution on shallow water at Blyth (Plate IX.)

Mr. Scott has also described a method of making the bottom of the structure flexible, so as to admit of its resting fairly on the ground. The floor beams are for this purpose made in pieces of small length, being placed single and double alternately, and connected together with bolts so as to adapt themselves approximately to the irregularities of the bottom on which they are to rest.

Available Capacity of Harbours of Refuge.—Mr. Mather remarks * that smaller vessels require 40 fathoms, while ships of 200, 300, and 400 tons require 60 to 80 fathoms, and giving room to swing clear of anchors, 150 vessels would fill up 470 acres of 18 feet and upwards, and 325 of 12 feet and upwards.

This would give as a mean 5·3 *vessels per acre.*

Minard allows for large merchant vessels one cable-length, which would give *about* 4 *vessels per acre.*

Captain Calver allows *three vessels* per acre for a small sheltered harbour of refuge.

Available Capacity of Anchorages and Natural Bays.—At Cardiff Flats there were at one time 224 vessels beached as close to each other as they could well be, in an open roadstead, and occupying a space of 560 acres, at the rate of 4 *vessels per acre.*

* Ships and Gales, by J. Mather: London, 1858.

CHAPTER VII.

DESIGN OF PROFILE, ETC., FOR TIDAL HARBOURS.

Weight better than Strength—Lighthouse Masonry—Effect of Configuration of Rocks—Forces from outside of Masonry—Horizontal Force—Cavettos and String-Courses—Rise of Spray—Vertical Forces—Back Draught—Forces within Masonry—Piers of insufficient Width—Rarefaction of Air—Rock Foundation—Durability and Strength of Rocks—Profile with large Materials—Profile with small Materials—Properties of Rocks—Profile of Conservancy—Walls of Horizontal and Vertical Profile—Clay Foundation—Application of Principles.

HAVING considered the few facts which have been ascertained regarding the action of the waves in the open ocean, I shall now direct the reader's attention to their effects in shallow water. The undulations in deep water are chiefly *whole waves*, and regarded by many as being purely oscillatory, while those in shoal water are breaking waves, and therefore regarded by all as waves of translation. We have hitherto been considering outer breakwaters erected in deep water, and which are constantly exposed to the waves; we now turn to piers and sea-walls which are placed within the range of the breaking surf, and which are exposed to its force for a limited period only, being sometimes left nearly or altogether dry by the receding tide.

Stability better attained by Weight than by Strength of Materials.—In dealing, then, with waves which are by all admitted to exert a true percussive force, the question arises as to how this force may be best resisted—whether by opposing to it *dead weight*, or a comparatively light structure, the stability of which is dependent on strong fixtures connecting it with the bottom. I cannot do better than quote the following remarks on this subject by the late Mr. Alan Stevenson, which were made with reference to the stability of lighthouse towers, but which apply more or less to every work which is placed within reach of the waves :—

"A primary inquiry in regard to towers in an exposed situation, is the question whether their stability should depend upon their *strength* or their *weight;* or, in other words, on their *cohesion* or their *inertia?* In preferring weight to strength we more closely follow the course pointed out by the analogy of nature, and this must not be regarded as a mere notional advantage, for the more close the analogy between nature and our works, the less difficulty we shall experience in passing from nature to art, and the more directly will our observations on natural phenomena bear upon the artificial project. If, for example, we make a series of observations on the force of the sea as exerted on masses of rock, and endeavour to draw from these observations some conclusions as to the amount and direction of that force as exhibited by the masses of rock, which resist it successfully, and the form which these masses assume, we shall pass naturally to the determination of the mass and form of a building which may be capable of opposing similar forces, as we conclude with some reason that the mass and form of the natural rock are exponents of the amount and direction of the forces they have so long continued to resist. It will readily be perceived that we are in a very different, and less advantageous position, when we attempt, from such observations of natural phenomena in which *weight* is solely concerned, to deduce the *strength* of an artificial fabric capable of resisting the same forces, for we must at once pass from one category to another, and endeavour to determine the strength of a comparatively light object which shall be able to sustain the same shock which we know by direct experience may be resisted by a given weight. Another very obvious reason why we should prefer mass and weight to strength as a source of stability is, that the effect of mere inertia is constant and unchangeable in its nature, while the strength which results even from the most judiciously disposed and well executed fixtures of a comparatively light fabric, is constantly subject to be impaired by the loosening of such fixtures, occasioned by the almost incessant tremor to which structures of this kind must be subject from the beating of the waves. Mass, therefore, seems to be a source of stability, the effect of which is at once apprehended by the mind as more in harmony with the conservative principles of nature, and unquestionably less liable to be deteriorated than the strength which depends upon the careful proportion and adjustment of parts."*

Movement of Lighthouse Masonry.—Although there is a great difference between the action of the sea on the masonry of harbours and of lighthouses, yet in both cases we have the same agent to deal with, and we know that if a given diameter of tower prove insufficient for the perfect masonry of a lighthouse, the breadth corresponding to that diameter must *a fortiori* be insufficient for a harbour with a similar exposure. In briefly adverting to some peculiarities of one or two of these structures, we shall farther learn how difficult it is to arrive at a correct appreciation of the exposure of different localities.

In November 1817, the waves of the German Ocean overturned, just after it had been finished, the Carr Rock Beacon,

* Account of the Skerryvore Lighthouse. By Alan Stevenson, LL.B., F.R.S.E. Edinburgh, 1848.

a column of freestone 36 feet high and 17 feet at the base, which was the largest size that the rock would admit of. The diameter at the level of high water was 11 feet 6, and at the plane of fracture 12 feet 9.

Modifying Influence of the Configuration of the Rock on Breaking Waves.—When the history of this beacon is contrasted with the records which have been preserved of the extraordinary structures that were erected on the Eddystone by Winstanley, we cannot help suspecting that that rock must be of peculiar configuration, by which the force of the waves against the buildings which have been successively erected on it, is modified, or to some extent diverted. The elaborate, though unsuitable design of Winstanley (*vide* Plate XII.) withstood the assaults of no fewer than eight winters. As the additions which he made from time to time to his original tower were in many instances anything but improvements, it may be questioned whether its fate (which was the result of an almost preternatural storm) was not accelerated by those injudicious alterations. His first year's work was a tower 12 feet high and only 14 feet diameter at the base, yet this stood during the whole of the first winter. The next year it was increased to 16 feet at the rock, a *polygonal* form being adopted, with an open gallery and vane with large ornamental scrolls, in all a height of 80 feet. This also stood for a winter. The next year it was increased to 20 feet at the rock by an outer ring of masonry, and the extreme height was raised to 120 feet. The open gallery and polygonal form were still retained; and numerous obstructive stages, a bluff projecting house for fishing, square chimneys, and colossal scrolls, were added. In spite of its great leverage and the extent of its surface, this uncouth structure, above which the sea was said to rise more than one hundred feet, stood, strange to say, for other four years, and disappeared only in November of the fifth winter, during one of the greatest storms ever experienced in this country, and of which I have already given an account in Chapter II.—from the records of which I think it may fairly be questioned, whether Winstanley's work was really knocked down by the sea, or was overturned by the violence of the wind.* The next lighthouse was Rudyerds, which at the level of the top of the rock was about 22 feet diameter, with a height of about 80 feet, and although consisting wholly of timber, stood forty-six years, when it was unfortunately destroyed by fire. It is farther remarkable that, while at the building of the Bell Rock, three stones, all dovetailed, wedged, and trenailed, were lifted after they had been permanently set, and at the Carr Rock Beacon, twenty-two stones were displaced, there was no instance of any but loose, unset blocks having been moved during the erection of the Eddystone, although the stones were of very similar weight, and fixed in much the same manner. Smeaton mentions that "after a stone was thus fixed we never in fact had an instance of its having been stirred by any action of the sea whatever."† During the third winter at the Carr Rock, nine stones, half a ton in weight, which were dovetailed, wedged, and trenailed, were removed *at the level of low-water neaps.* The trenails were all broken. *At* 2½ *feet above the mean level of the sea* five stones were in like manner moved, and during the fourth winter seven other stones were moved at a height of *three feet above the high-water level.* There seems, therefore, good reason for believing that the form of the Eddystone rock shelters to some extent the structure that rests upon it. I may also mention that my

* An Historical Narrative of the Great and Tremendous Storm which happened Nov. 26, 1703. Lond. 1769. P. 148.

† Account of the Eddystone, p. 132.

friend Captain Fraser, who was lately engaged in the arduous work of erecting a lighthouse on the Alguada reef in India, entertains somewhat similar views regarding the Eddystone rock.

Additional and very remarkable corroboration of these views has also been recently afforded by the experience derived in the course of construction of the Dhuheartach lighthouse, which was finished in 1872, after having occupied five working seasons.

The Dhuheartach rock lies about 15 miles to the W.S.W. of Iona in Argyleshire. It is 220 feet long and about 30 feet high, the tower being raised to the height of 130 feet above the sea. The rock is everywhere surrounded by deep water, and it is of an elliptical form. During a summer gale *fourteen* stones, each of two tons, which had been fixed in the tower by joggles and Portland cement at the level of 37 feet above high water, were torn out and swept off the rock into deep water. (*Vide* Plate XIII.)

Now it is a remarkable fact that the level above the sea at which these blocks were removed by a summer gale is the same as that of the *glass panes* in the lantern of Winstanley's first lighthouse, which nevertheless stood successfully through a whole winter's storms. And in the tower, as last constructed by Winstanley, there was an open gallery at the same level as the former lantern, above which the cupola and lantern were supported, and which stood for four winters. In other words, at the same level at which thin panes of crown glass stood successfully for a winter at the Eddystone, and at which for other four winters the open gallery with closed-in cupola above stood without damage, the fourteen joggled stones of two tons each were swept away at Dhuheartach.

The conclusion, then, which is fairly deducible from these

facts is that *the impact against a lighthouse depends upon the relation subsisting between the height of the waves at the place and the height and configuration of the rock above and below low water, and perhaps also on the configuration of the bottom of the sea at the place.*

Thus, while the rock at Dhuheartach, from its height above the waves, forms a protection against the smaller class of waves, it operates as a dangerous conductor to the largest waves, enabling them to exert a powerful action at a much higher level than they would attain had the rock been lower. Hence the fact that the highest levels at which set stones were moved at the Carr Rock was 3 feet above high water, and at Dhuheartach 37 feet above high water, may be accounted for by the different configurations of the rocks, without assuming that the waves are exceptionally high at Dhuheartach. Plate XIII. will be found to exhibit an interesting view of these and other similar lighthouses, all laid down to the same scale.

It is of great importance that these facts should be kept in view, and that the Eddystone should not be regarded as a safe model for imitation at all rocks which are exposed to a heavy sea.

Different Forces which assail the Masonry of Harbour Works from the outside.—The determination of the stability of a practically monolithic mass, such as the Eddystone or Bell Rock Lighthouse, though difficult enough, is, however, of a simpler nature than that of the disconnected materials of a harbour work. In the masonry of lighthouses the sea is excluded from the joints of the stones, but in most harbour works the jet of water enters freely into the interior of the masonry, and introduces such complexity into the question as to render it impossible, at least in the present imperfect state of our information, to give any rules for directing the engineer. I can only, therefore, simply point out the nature of the different forces which enter into the question of the stability of such loosely built structures, so as to show in what directions the sea attacks the work. The impact of the waves against the outside of a sea-wall or pier gives rise to four distinct forces, namely—1st, The direct horizontal force which tends to shake loose, or drive in, the blocks of which the masonry consists. 2d, The vertical force acting upwards on any projecting stone or protuberance, as well as against the lying beds of the stones. 3d, The vertical force acting downwards, which results either from the wave breaking upon the toe of a talus wall, or from its passing over the parapet, and falling upon the pitching behind, so as to plough it up. 4th, The *back-draught*, which tends by reaction from the wall to remove the soft bottom, and in this way to undermine the lower courses of the work.

It may be concluded from the above that the points which require to be most attended to are—the contour and quality of masonry of the wall itself—the parapet, which, if not of proper form, or of insufficient height, leads to damage in the pitching behind it ; and lastly, the foundation-courses, in the design and construction of which, if similar precautions be not attended to, underwashing of the bottom may in some situations take place, so as to leave the lowest courses without protection.

1. *Horizontal Force against Sea-Walls—Flush Dynamometer—Observations on Force at different heights.*—In 1858 I made some experiments at Dunbar, which were continued till the completion of the new wall there, in order to ascertain the relative forces exerted at different levels. The first series began with the old curved sea-wall (Fig. 15). For this pur-

pose it was necessary to sink the dynamometers into the

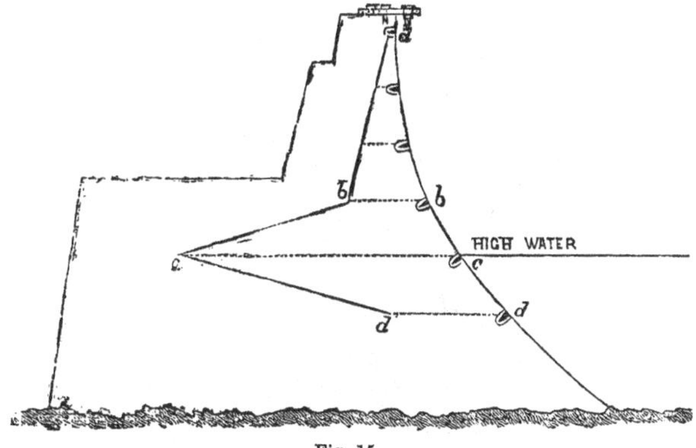

Fig. 15.

masonry, so that their discs should be nearly *flush* with the line of wall. As Figs. 16, 17 represent those flush dynamo-

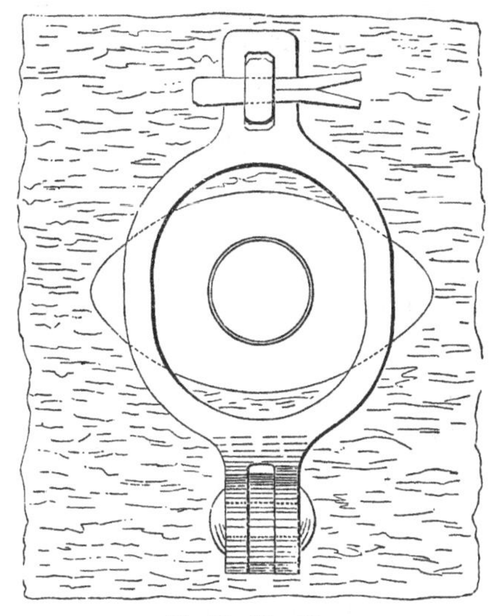

Fig. 16. Elevation.

meters both in elevation and section, no description of them seems necessary. The holes *a a a*, Fig. 17, shown in the under

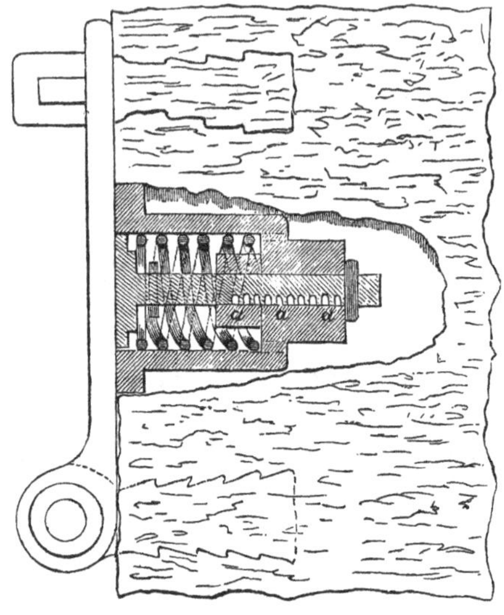

Fig. 17.

side of the rod, were made for holding shot which was expected to drop out when the rod was pushed so far out as to leave them unsupported. This arrangement, which promised to be a good check on the indications of the leathern index, was not, however, found to answer, but my friend Mr. Alan Brebner, C.E., has suggested that a cylinder of wax might indicate well by having its surface scratched by a needle attached to the movable end of the spring. The cavities in the stone into which the flush dynamometers were sunk were made larger on the upper side than was necessary for holding them, in order to form at the top a reservoir for air, the compression of which on the impact of the wave against the disc permitted it to pass inwards on the stroke of the wave. Owing to a most unfortunate uncertainty regarding the readings of some of the instruments employed in these observations, and which was only discovered when the results, extending over a period of years, were examined for reduction, it would not be warrantable to deduce from them any rule or formula. The source of error having been now discovered, other results, on which full reliance can be placed, will, it is hoped, be obtained; but the lines *b b*, *c c*, *d d* (Fig. 15), may be held to represent generally the nature and directions of the forces.

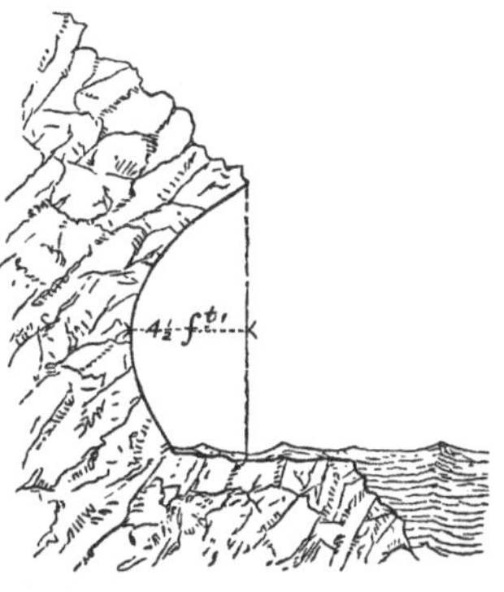

Fig. 18.

Before leaving this part of the subject I may refer to the action of the waves of the tideless Mediterranean. When in Italy in 1864 I made the accompanying sketch (Fig. 18), which shows the scooping out of the solid rock during the lapse of an unknown number of centuries on the coast between Mentone and Ventimiglia, and which illustrates well the sudden reduction of force which it is known takes place immediately below the water level.

In 1838 I made several cross sections of the forms which the waves scooped out of the clay banks of the Bristol Channel at Cardiff moors. Fig. 19 represents the general form of profile.

2. *Observations on Vertical Force.*—Simultaneously with the other observations at Dunbar, two marine dynamometers, of the common construction (*a*, Fig. 15), were fixed to a piece of wood which was bolted to the top of the cope of parapet. The instruments were fixed with their discs projecting over the edge of the cope, and pointing downwards so as to ascertain

the upward force of the ascending body of water and spray. The *maximum vertical force* recorded at the cope of the wall,

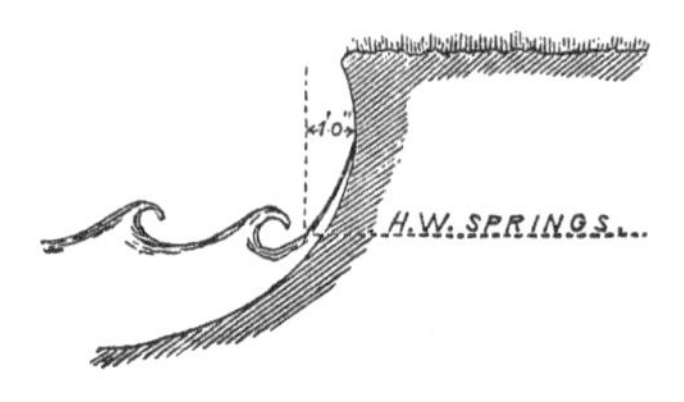

Fig. 19.

a, Fig. 15 (23 feet above the sea at the time that the observation was made), was upwards of one ton (2352 lbs.) per superficial foot, while the greatest *horizontal* force recorded by the highest of the *flush* dynamometers, which was fixed 18 inches lower in the stone immediately below the cope, never exceeded 28 lbs. The great vertical force at and near the top may at first sight appear to be somewhat anomalous, but it must be recollected that the discs of the dynamometers were in the one case parallel, and in the other opposite, to the line of direction of the particles, as altered by the wall. *The vertical force tending to raise the cope of this sea-wall is therefore about* 84 *times greater than the horizontal force tending to thrust it inwards.* This shows that the higher the spray is allowed to rise the less force will be exerted horizontally against the masonry near the top, unless any part of it projects beyond the face of the wall. Therefore, to make what may be called an easy wall for the sea, the outer edge of the cope should be slightly rounded, or the stone itself should be set an inch or two back from the face of the wall.

Danger from Cavettos and String-courses on Sea-Walls.—It follows further from these observations that not only should all overhanging string-courses be avoided, but that even very rough ashlar stones with large protuberances on the seaward face are undesirable. At the harbour of Stonehaven (Fig. 20)

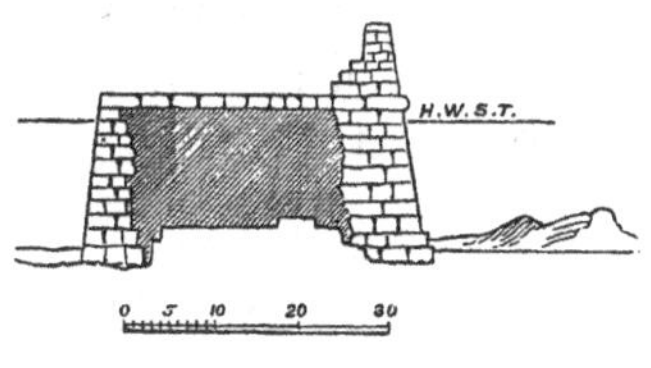

Fig. 20.

a "string" or "bottle" course had to be hewn off, in order to prevent the concussions which took place during storms, and which were so great as to shake the superincumbent masonry. Even such small objects as an upright pole have been found, from their leverage, by their catching the spray, to disturb the masonry to which they were fixed. This was proved at Cockenzie, East Lothian, where a flagstaff had to be removed on account of its shaking the parapet when the waves struck it.

Observations of Rise of Spray on Vertical and Curved Walls.—In connection with the Dunbar experiments, it may be added that the spray—on an average of seventeen observations taken roughly—*rose on the hollow curved wall about seven times higher than the waves which projected it, and on the vertical wall,* taking a mean of twenty-three observations, *it rose* 6.6 *times higher.*

3. *Vertical Force acting Downwards—Damage to Roadway.*—The water, after striking upon the sea-wall, rises in large volumes above the parapet, and, descending with great force, dashes against the pitching of the roadway. If the pitching stones are not sufficiently heavy and closely assembled, they may be ploughed up by the descending water, when the interior of the work is at once laid open to the destroying

element. The late Mr. James Bremner, of Wick,* thus describes the disastrous consequences which resulted from the breaking up of the roadway pitching of the old harbour of Pulteneytown, and the manner in which he protected it from after damage. "By the 20th of September in that year" (1827) "the whole operations would have been substantially completed, but, on the 10th of that month, a violent storm arose, during which, notwithstanding of precautionary means, which he had always provided against ordinary storms, by blocking up and forming substantial obstructions against the sea, for temporary defence, an extent of building of not less than 100 feet of the breakwater-head was, during one tide, swept away, and reduced to the level of the water."

"To prevent the recurrence of such misfortunes, the author reared a wall, of large rough stones, under the parapet; he compacted the roadway pitching with fir-wood wedges, on which cills 1½ inch broad, running along the roadway, were spiked down at distances 10 inches apart. On these cills, boards of an inch thick were fixed, and closely joined together, the outer ends lying to the foot of the parapet, while the inner ends reached half-way over the coping of the front wall, so that the sea, falling over the parapet, was not permitted to touch the pitching."

4. *Horizontal Force of the Back Draught of the Wave.*— Dynamometers having their discs *facing the wall* were fixed to a pile placed immediately outside of the foundation of the Dunbar bulwark, while others were fixed to the same pile with their discs pointing seawards. On one occasion, owing no doubt to the concentration of the watery filaments by the sea-wall, the force of recoil was equal to one ton, while the direct

* See Treatise on the Planning and Building of Harbours, etc. By James Bremner, M.I.C.E. Wick, 1845.

force of the waves before they had reached the wall was only 7 cwt per square foot, or *the force of recoil was equal to three times the direct force.*

Destructive Forces acting within the Masonry.—Within the masonry, as well as without, the waves exert force in the following different ways:—1st, By the propagation of vibrations produced by the shock of the waves on the outer or sea-wall, through the body of the pier to the inner or quay-wall; 2d, by the direct communication of the impulses through the particles of the fluid occupying the interstices of the hearting, so as to act against the back joints of the face stones of the quay; 3d, by the sudden condensations and expansions of the air in the hearting, so as to loosen, and at last to blow out, the face stones of the quay, combined with, 4th, the hydrostatic pressure of the water, which is forced through the sea-wall, and, from want of free exit, is retained and acts as a *head* at the back of the quay, and which, however small in quantity, will, as in a Bramah press, act upon all surfaces exposed to its pressure, however great those surfaces may be. The three last causes are probably the most efficient agents in the work of destruction.

Examples of Piers of Insufficient Width destroyed by Forces acting within the Masonry.—Although, with the single exception of Wick breakwater, there is no instance on record of any pier being overturned by one stroke of the waves, and it may generally be concluded that, in all ordinary exposures, harbour works capable of affording the accommodation required for shipping will necessarily possess sufficient mass to resist destruction in this wholesale manner, yet many instances have occurred of large portions of masonry being moved *en masse* by a single stroke of the sea. As an example occurring in a comparatively sheltered locality, we may adduce the harbour of Millport in the Isle of Cumbrae. In December 1862, Mr. Jamieson, the harbour master, was on the pier when a very heavy wave struck the parapet with most unusual violence, throwing up a jet of water which he described as "half spray, half solid," to the height of 10 or 12 feet above it. The wave covered that part of the roadway at the outer end which had not the protection of a parapet to the depth of about 18 inches, and moved (not rolled) a mass of sheet lead weighing 12 cwt. along its surface, until it came in contact with one of the mooring pauls, which arrested its motion. On examining the masonry, which he did as soon as the waves would permit, he found that a large portion of the parapet had been thrown back.

While the parapet remained in the same state in which the storm left it, I had an opportunity of measuring it with Mr. Jamieson, and found that 33 feet in length had been bodily thrust backwards to the maximum extent in the middle of 4 inches. The level of the bottom of the parapet was, according to Mr. Jamieson, about 7 feet above ordinary springs, and I found it to be 6 feet 3 inches above the level of the *lepas* or barnacle shell. The masonry, which is freestone, was all joggled with cubes of greenstone, and there were side straps on the cope. The parapet, 5 feet 5 inches in height, consisted of five courses including the cope, and Mr. Jamieson said it was nearly of solid ashlar, there being hardly any rubble hearting.

It sometimes happens that from the forces acting within the face-walls to which we have referred, serious damage has resulted if the pier be of insufficient width. One case of this occurred at an exposed port on the east coast of Britain. Without entering upon the details of this work, it is sufficient to mention that the pier, which had a sea-wall curved vertically to a radius of 32 feet, received much injury during the execution of the work. As the damage gave rise to certain legal questions between the contractor and the promoters, an investigation was made with reference to those questions, when the following interesting information was obtained. It was found, after the storm which occasioned the greatest damage, that of 230 feet of finished pier, about 120 feet had been demolished, and the materials formed a heap, the seaward face of which had assumed a natural slope or angle of repose of about 4 to 1. After a minute inspection of the *parapet* wall of that part of the pier that had been left standing, not the smallest appearance of failure could be anywhere detected. The *sea-wall* was in like manner unaffected, with the exception of one or two points where some appearance of starting or very slight shifting was noticed. The *pitching* of the roadway was also quite sound and entire. But on turning to the sheltered side of the pier forming the *quay*, the work presented a very different aspect. This wall was very much shattered for a distance of about 140 feet. Most of the stones in the facework were skirted or cracked at the corners, and had evidently been moved in their beds to a greater or less extent, and some of them, at and near the middle courses or half-way up the wall, had been thrust out as much as $\frac{7}{8}$ of an inch beyond the face line. It was found, on examining the heap of debris of that part of the work which had been thrown down, that the lower courses of the quay-wall for about six feet above the foundation had not been overthrown, though they were much shattered, and the general line of the masonry was also greatly distorted. The cause of the damage, in the opinion of those resident at the place, was want of strength in the *quay-wall* to resist the forces which were propagated through the pier from the outside. The truth of this statement was fully borne out by the fact that the damage occasioned by *five* different storms, had in every instance made its first appearance on this *inner* wall. The breadth of the pier, at the level of the roadway, was 26 feet 4 inches, and at the level of high water, 28 feet.

Another case which also gave rise to legal questions was that of a harbour situated also on the east coast of Britain, which, although open to the N.E. swell, was nevertheless to some extent protected by a considerable reef of rocks extending in front of it. It differed from the former case in being of very bad workmanship. The stones composing the face of the sea-wall were imperfectly dressed, being all "*lean to the square.*" The hearting had been carelessly thrown in, without hand packing, and the quay-wall was also very ill dressed, and from an inspection of the work after the accident, there did not seem to have been any regular *backing.* While walking along the roadway of the pier, before the storm took

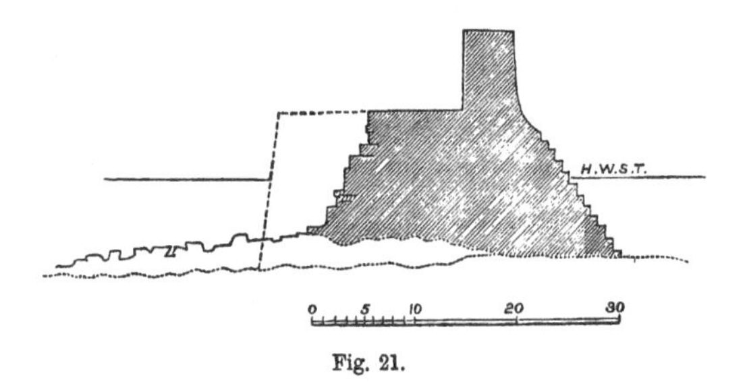

Fig. 21.

place which did the damage, I was surprised on seeing jets of air and water suddenly projected into the basin of the harbour with a loud report. These were found to proceed from the stroke of the waves on the sea-wall outside, which was transmitted through that wall, and through the hearting and quay-wall. The waves at the time did not much exceed 3 or 4 feet in height, and yet the impulse of almost every one was propagated through the pier as if through wicker-work, making its presence evident by the jets of air and water, some of which extended into the basin of the harbour to probably 20 feet beyond the front of the quay. The width of the pier (Fig. 21) was 24 feet at the level of the roadway. During a severe storm from the E.N.E. which took place some time after the work was finished, the greater portion of this work was thrown down, and it was observed that the *quay-wall was at some places the only part that gave way.* The spray during the gale rose about 20 feet above the top of the parapet. A sailor resident at the place stated that during the gale, but before any damage had taken place, he repeatedly saw water spouting up through the pitching (close to the cope of the quay-wall) to the height of about 16 feet above the roadway. This jet would of itself indicate a pressure against the back of the quay, even as high up the wall as the cope itself, of about half a ton on the square foot.

Another instance of a similar kind was seen at a harbour on the west coast of Britain (Fig. 22), where the length of fetch

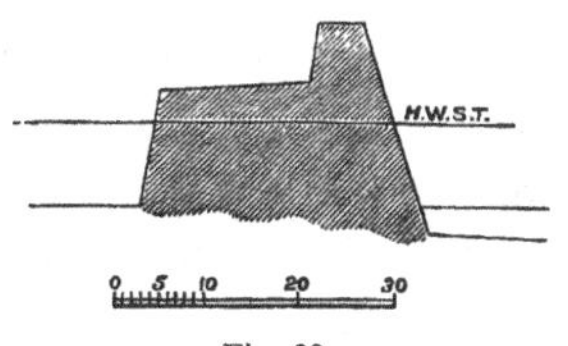

Fig. 22.

was only 40 miles. The breadth at level of high water was 26 feet, and the workmanship seemed to be of good quality. In this case, the pitching was the first part that was seen to give way by those resident on the spot. It was forced upwards by the air and water in the interior. Ultimately the sea and the quay walls were also much damaged.

Minimum Width of Piers in German Ocean.—After comparing different piers in the German Ocean where we have most examples, and from careful consideration of the other data, I should not feel disposed to recommend that vertical piers, fully exposed to the *ordinary* waves of that sea, should be less than from 35 to 45 feet broad at the level of high water. But of course this remark does not apply to anomalous cases where the sea is exceptionally high, but, as stated, to works exposed to the ordinary waves.

Examples of Damage from Rarefaction of the Air outside of the Masonry.—Outer sea-walls, as well as the inner quay-walls, are also liable to damage from pneumatic action *ab infra*. The sudden "*back draught*" occasioned by the relapse of a heavy wave after it has broken against a building, produces, according to the late Mr. Walker,* a certain amount of rarefaction in the surrounding air. Mr. Walker seems to have been the first to notice the existence of this phenomenon, which was manifested in a remarkable and extraordinary manner at the Eddystone Lighthouse during a heavy sea in the year 1840. On that occasion, the entrance door of the tower, which was made secure by strong bolts against the force *ab extra*, was driven outwards by a pressure acting *from within the tower*. The strong bolts and hinges are stated to have been broken. This interesting and almost incredible incident seems to prove that by the instantaneous sinking of the wave the atmospheric pressure was suddenly, and to a very considerable extent, removed from the air on the outside of the door. As the air within the tower continued still to receive its usual pressure, which would be communicated freely from the windows and lantern door in the *upper* parts of the building, it appears suddenly to have burst the door outwards in order to supply the partial vacuum, and thus to

* Inst. Civ. Eng. vol. i. p. 115.

restore between the air outside and inside of the building the equilibrium which had been disturbed by the sudden subsidence of the wave. It is perhaps from this, among other causes, that stones sometimes start out of the facework of sea-walls in a very unexpected manner, and if not speedily noticed, soon project so far as to allow the waves to remove them altogether, when the worst consequences may ensue.

The accompanying diagram, Fig. 23, represents a case of this which occurred at the harbour of Buckie, in Banffshire,

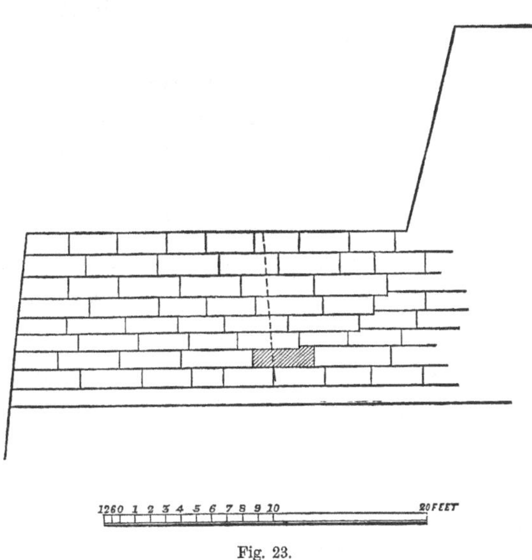

Fig. 23.

in which the combined effects of condensation and back draught started a stone from its bed in the face of the end of the pier. The whole of the masonry had been finished several months before, and had encountered successfully two or three very hard gales, when it was noticed, after the occurrence of another heavy sea, that the stone, which is hatched black in the diagram, had started from its bed, and was projecting several inches out from the wall. The stone, which was a stretcher, 4 feet long and 14 inches thick, had no doubt been originally set slack in the work. A heavy wave had first driven water and condensed air through the joints among the backing, and then the sudden subsidence of the wave had produced a rarefaction, and lastly the head of water and the condensed air in the inside of the work had, in bursting outwards, started the stone from its bed. It was made secure by a vertical bolt driven from the coping course in the direction shown by the dotted line. It appears probable, then, that the stones composing the masonry of sea-works may be subjected to simultaneous expansions of the condensed air behind, and rarefactions occasioned by the collapse of the wave in front.

Buildings on a Rock Foundation—Selection of Stones.—Having now adverted to the different forces to which marine works are exposed, I shall next consider what kind of design is most suitable where the bottom consists of hard rock. Such a foundation will render unnecessary any precautions arising from the wasting of the bottom, and, *cæteris paribus*, there does not seem to be any reason for preferring a talus to a vertical wall.

The question of preference where the rock is sound, will in the main depend upon the kind of material which can be obtained. Should the stone be scarce or costly, and the quality such as to warrant the introduction of masonry of the best description, the vertical wall may be found to be the most economical. Where freestone is to be used, it is not only desirable that it should be got in large blocks, but that the face stones should possess considerable hardness. All materials which are subject to decay from atmospheric influence should either be entirely rejected or reserved for the inner face of the parapet. Special care is also necessary in selecting the stones for the lower courses, and all the more so, if the beach consist of hard gravel or boulders. Where these occur on an exposed coast, granite or other hard stone should always be preferred. There are beaches where even blocks of greenstone waste away with alarming rapidity.

Cessart found timber to resist abrasion from rolling gravel ten times longer than limestone. The timber planking which he used was fixed *vertically* on the masonry.*

In 1870 I noticed a somewhat remarkable fact at the foundations of the old castle at Sandgate, near Folkestone. The granite blocks forming the outer scarcements had been bound together with iron bars, secured with lead. The lead had, owing no doubt to slower oxidation, held its ground against the sea better than either the iron or the granite, and stood up about $\frac{1}{8}$ inch above the surface of the more rapidly wasting materials.

Relative Durability of different Rocks.—The superiority of granite to greenstone is proved by the following experiments on the times required to abrade $\frac{1}{16}$ inch of each kind of stone. They were made with the same weights and grinding agents, and with equal cubes of each material:—

30 minutes were required for Queensferry (Carlin Nose) greenstone.
40 ,, for greenstone from Barnton, near Edinburgh.
60 ,, for Peterhead granite.

Mr. Murray of Sunderland also established, many years ago, the superior power of resistance possessed by granite over greenstone.

* Ouvrages Hydraul. p. 83.

Strength and Hardness of Materials.—At Anstruther harbour the freestone of the district was, from motives of economy, employed for the sea-wall, in consequence of the failure of the rock from the excavations. Stones thrown up by the sea caused fractures of the face-stones, which gradually falling out produced extensive damage to the pier itself.

Wasting of Rock Foundations.—Rocky ledges often break up at the foundation of an exposed sea-wall, owing to the heavy fall of water in front of the work. An occasional examination should, from time to time, be made, to ascertain whether this dangerous action is taking place, as it may lead at last to undermining of the wall.

Profile when Materials are Large and Unworkable.—If the materials are abundant, but of an unworkable nature, a long talus wall will be found most economical. For such walls the rate of slope must depend very much upon the exposure of the place, and upon the plentifulness of rubble stone hearting. The easily-dressed and naturally flat-bedded materials, which the stratified rocks of the secondary formation very often furnish, are especially applicable to the construction of vertical walls ; while the uncouth blocks of the primary and igneous formations are better suited for talus walls. Such rocks as gneiss, the schists, basalts, greenstones, porphyries, and the tougher kinds of granite, are best fitted for this purpose. With some of these rocks, the angularity of the pieces, and the excessive difficulty and uncertainty of dressing, render it necessary to assemble them without almost any alteration of their shape, so as, by an adaptation of their salient and re-entrant angles, to make a kind of random rubble face-work. In this kind of work mortar was formerly very seldom employed, but Mr. Hartley first, and afterwards, at Holyhead, Sir J. Hawkshaw successfully introduced rubble, consisting of enormous unsquared masses of rock set in hydraulic mortar. The *parapet*, which usually consists of squared masonry on the outside, surmounted by a heavy cope, used generally to be the only part of sea-works which was set in lime-mortar, the joints being pointed or lipped with Roman or Portland cement. The valuable qualities of the limestone from Halkin Mountain in North Wales, Aberthaw in Glamorganshire, and more especially Arden in Ayrshire, are generally known, and render them well suited for building parapets, but Portland and Zumaya cements are of all mortars undoubtedly the best.

Profile when Materials are of Small Size—Cycloidal Curve. —Where the materials are light and of small sizes, it is desirable to design the work so as to equalise the action of the sea over the whole work, and not to concentrate it against

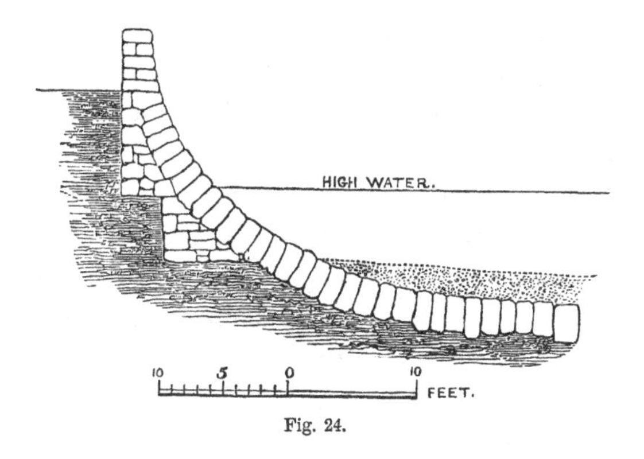

Fig. 24.

any particular place. Mr. Russell states that the cycloidal form was recommended for this purpose by Franz Gerstner of Bohemia. The only instance of the adoption of this curve with which I am acquainted (Fig. 24) is in a sea-wall which was erected at Trinity, near Edinburgh, by the late Mr. Robert Stevenson, in 1822.

Table of Properties of different Rocks.—In judging of the qualities of different quarries for harbour purposes, the importance of a high specific gravity should not be overlooked. The accompanying table is useful in showing this. When s represents the specific gravity, n the number of cubic feet in ton in air, n the number of cubic feet in sea water, a side of cube weighing a ton in water, and w weight of a ton-block in water—

$$n = \frac{2240}{s \times 62.5} \qquad n' = \frac{2240}{(s \times 62.5) - 64.25}$$

$$a = \sqrt[3]{\frac{2240}{(s \times 62.5) - 64.25}} \qquad w = \frac{2240 - 64.25\, n}{2240}$$

The specific gravities of the stones marked thus * are given by the Parliamentary Commission which was appointed to report on the best stone for the new Houses of Parliament, and the specific gravity of Beton is assumed on the authority of Minard in his *Cours de Construction.* All the others I took by means of a spring balance, a method which I strongly recommend, from the quickness as well as accuracy with which the specific gravities are obtained. Any piece of rock, however rough and unsymmetrical, may be readily suspended from the balance, and the reading on the scale at once noted. The stone is next immersed in water, and its weight again noted, the whole process of getting the weights in air and water not occupying so much as one minute. Dressing and weighing a cubic foot of stone, which is so common with quarry masters, is both an inaccurate and tedious method, which need never be resorted to.

TABLE OF PROPERTIES OF DIFFERENT KINDS OF ROCK.

NAMES OF ROCKS.	s	$s \times 62.5$	$\frac{2240}{s \times 62.5} = n$	$\frac{2240}{(s \times 62\frac{1}{2}) - 64.25} = n$	$\sqrt[3]{\frac{2240}{(s \times 62.5) - 64\frac{1}{4}}}$	$\frac{2240 - 64\frac{1}{4} n}{2240}$
	Spec. Grav.	Weight of a Cubic Foot in Air.	No. of Feet to a Ton in Air.	No. of Feet to a Ton in Sea Water, Sp. Gr. 1.028.	Side of Cube corresponding to Last Column.	Weight of a Ton Block in Sea Water, Sp. Gr. 1.028.
		Lbs.	Cubic feet.	Cubic feet.	Lineal feet	Ton.
Basalt (porphyritic) ...	2.99	186.87	11.9	18.26	2.63	.658
Greenstone	2.92	182.50	12.2	19.00	2.66	.650
Syenite	2.91	181.9	12.3	19.04	2.67	.647
Clay-slate	2.90	181.25	12.3	19.15	2.67	.647
Mica-schist	2.89	120.6	12.4	19.24	2.68	.644
Gneiss	2.82	176.2	12.7	20.00	2.71	.635
Amygdaloidal greenstone	2.75	171.87	13.0	20.81	2.75	.627
Chlorite-schist	2.74	171.25	13.0	20.93	2.75	.627
Greywacke	2.73	170.62	13.1	21.06	2.76	.624
Clinkstone	2.73	170.62	13.1	21.05	2.76	.624
Red granite	2.71	169.37	13.2	21.30	2.77	.621
Slate (old red sandstone formation) ...	2.71	169.37	13.2	21.30	2.77	.621
Chalk	2.70	168.75	13.2	21.43	2.77	.621
White primary limestone (marble) ...	2.70	168.75	13.2	21.43	2.77	.621
Red primary limestone (marble), with crystals of augite embedded	2.66	166.25	13.4	21.96	2.80	.615
Chlorite	2.64	165.00	13.5	22.23	2.81	.612
Granular quartz rock ...	2.63	164.37	13.6	22.37	2.82	.610
Grey granite	2.61	163.12	13.6	22.65	2.83	.610
Flinty slate	2.57	160.62	13.9	23.24	2.85	.601
Red felspar porphyry ...	2.55	159.37	14.0	23.55	2.86	.598
Pitchstone	2.53	158.12	14.1	23.86	2.88	.595
Serpentine	2.46	153.75	14.5	25.02	2.92	.584
Compact felspar	2.45	153.12	14.6	25.20	2.93	.581
Sandstone	2.41	150.62	14.8	26.00	2.96	.575
Roestone (oolite)... ...	2.36	147.50	15.1	26.88	2.99	.567
Beton	2.2	137.05	16.3	30.6	3.13	.532
Magnesian limestone * ...	2.18	136.25	16.4	31.11	3.14	.529
Oolite *	2.05	128.12	17.4	35.07	3.27	.501
Parrot coal (American) ...	1.54	96.25	23.3	70.00	4.12	.331
Bituminous shale ...	1.50	93.75	23.8	75.93	4 23	.317
Cannel coal	1.24	77.50	28.9	169.05	5.53	.171

NOTE.—This table is calculated for sea water of the specific gravity of 1.028 or 64¼ lbs. to the cubic foot.

The following Table, showing the pressure of sea water per superficial foot for different depths, will be found useful in connection with that given above :—

TABLE showing Pressure of Sea Water per Superficial Foot for Different Heads above Centre of Gravity of Surface exposed. Specific Gravity taken at 1.028.

Height in Feet.	Pressure in lbs. per Sq. Foot.	Height in Feet.	Pressure in lbs. per Sq. Foot.	Height in Feet.	Pressure in lbs. per Sq. Foot.	Height in Feet.	Pressure in lbs. per Sq. Foot.
1	64¼	26	1670½	51	3276¾	76	4883
2	128½	27	1734¾	52	3341	77	4947¼
3	192¾	28	1799	53	3405¼	78	5011½
4	257	29	1863¼	54	3469½	79	5075¾
5	321¼	30	1927½	55	3533¾	80	5140
6	385½	31	1991¾	56	3598	81	5204¼
7	449¾	32	2056	57	3662¼	82	5268½
8	514	33	2120¼	58	3726½	83	5332¾
9	578¼	34	2184½	59	3790¾	84	5397
10	642½	35	2248¾	60	3855	85	5461¼
11	706¾	36	2313	61	3919¼	86	5525½
12	771	37	2377¼	62	3983½	87	5589¾
13	835¼	38	2441½	63	4047¾	88	5654
14	899½	39	2505¾	64	4112	89	5718¼
15	963¾	40	2570	65	4176¼	90	5782½
16	1028	41	2634¼	66	4240½	91	5846¾
17	1092¼	42	2698½	67	4304¾	92	5911
18	1156½	43	2762¾	68	4369	93	5975¼
19	1220¾	44	2827	69	4433¼	94	6039½
20	1285	45	2891¼	70	4497½	95	6103¾
21	1349¼	46	2955½	71	4561¾	96	6168
22	1413½	47	3019¾	72	4626	97	6232¼
23	1477¾	48	3084	73	4690¼	98	6296½
24	1542	49	3148¼	74	4754½	99	6360¾
25	1606¼	50	3212½	75	4818¾	100	6425

Buildings on Soft or Sandy Foundations.—It has been already shown that, irrespective of the quality of the masonry, the two points in the structure which are weak or dangerous are the *top* and *bottom* of the wall. With a hard rocky bottom, properly dressed, the risk of failure at the foundations is removed ; on the other hand, where the shore consists of soft or rotten rock, moving shingle, or sand, it is obvious that provision must be made against both those sources of evil. Indeed, if we consult the history of harbours, we shall find that the most frequent cause of damage is the corroding action of the waves reflected by the masonry against the shore.

Profile of Conservancy, and Underwashing.—The general slope of a fragmentary beach must depend upon the size and nature of the particles, and the force of the sea. The dissimilarity between its slopes near the levels of high and low water arises from a decrease in the force of the waves, caused by their being broken before they reach the high-water mark. The great object, therefore, is to design the profile of the wall, so as to alter as little as possible the symmetry of the beach. Where isolated rocks or large boulders are left projecting above the surface of a sandy shore, there will generally be formed around them hollows corresponding in depth and form to the kind of obstruction which the rocks present. The principal point in the design, therefore, is to avoid great and sudden obstructions to the movement of the water. The best form which could be adopted in any situation would, of course, be the contour of the beach itself; but this would answer no possible purpose; and, as the wall is to consist of heavy blocks of stone instead of minute particles of sand, it is clear that a much steeper slope may be adopted than that which we may call *the profile of conservancy* of the shore, provided the lower part of the slope be flattened out so as to meet the sand at a low angle. The action of a bulwark is to arrest the waves before they reach the general high-water mark, and to change the horizontal motion of the fluid particles to the vertical plane, or to compel the waves to destroy themselves on an artificial beach consisting of heavy stones. To prevent underwashing, the two following requisites should therefore be as far as possible secured:—1st, The foundation courses of the wall should rise at a very small angle with the beach, so that their top surfaces may be coincident with the profile of conservation of that portion of the beach out of which the wall springs; 2d, The outline of the wall should be such as to allow the wave to pass onwards without any sudden check till it shall have reached the strongest part of the wall, which should be placed as far from the foundation as possible.

Loose Rubble a good Protection for the Foundations, and for acting as a cover for Bulwarks for protecting Land against the Sea.—Loose blocks of angular rubble furnish, in most cases, the best possible security when the soil is soft or friable, for the waves are swallowed up by the interstices. A regular sloping sea-wall or bulwark, with a smooth surface, becomes, when the soil is soft, a *double-edged sword* in working its own destruction at top and bottom; for it transfers the duty of destroying the waves from the masonry to the unprotected soil at the top, and to the loose sand or gravel at the bottom, of the wall. While the foundations are underwashed by the reaction upon the soft bottom, the upper parts of the masonry are deprived of support by the falling water and spray, which are led up by the masonry, and soon wash away the soil at the top.

Walls of Horizontal Curvature for protecting the Coast-line.—But it must be further noticed that, if the wall be curved in the *horizontal* plane, or consist of kants inclined at an angle to each other, the foundations must be carried lower at the convex parts and at the salient angles, for at those parts there is a greatly increased scour. This concentration in the scouring action may be very satisfactorily seen at low water, where an isolated rock or a boulder of pyramidal form projects above the surface of a sandy sea-beach. Pools of greater or less depth will always be found at the angles of the boulder, while at intermediate parts the level of the sand is much higher.

Walls of Vertical Curvature.—For the reasons which have been stated, it is plain that a vertical wall is in most cases unsuitable for a sandy beach. Instead of altering the direction of the wave at a distance from its foundation, the whole change is produced at that very point; and, unless the wall be founded at a considerable depth, its destruction is all but certain. Where the materials are costly, but admit of being easily dressed, I am disposed to think that a horizontal, or nearly horizontal, apron or platform of timber or masonry, connected with a vertical wall by a quadrant of a circle of sufficient radius, may be found answerable. Such a form will prevent to a considerable extent the danger of reaction, by causing the alteration in the direction of the wave to take place at that part where the wall is strongest, and which is also at the greatest possible distance from the toe or curb course. If the materials are abundant, and of a rough nature, a cycloidal wall with vertical and horizontal tangents, somewhat similar to that erected at Trinity, already referred to, may be adopted with advantage. But a very serious objection to all forms of curved walls, unless the radius be large, is the weakness which results from the use of wedge-shaped face-stones. The impact of the sea on materials of that form may be compared to a blow directed upwards against the intrados of a stone arch—the direction of all others in which the voussoirs are most easily dislocated. This action can only be resisted by very careful workmanship in the dressing and setting of the *backing*. Another objection, applicable to all except tideless seas, such as the Mediterranean, arises from the varying level of the surface of the water; for what may be best at one time of the tide cannot be equally suitable at another.

Treacherous Nature of Clay as a Foundation—Ardentallan Pier.—A special caution regarding clayey bottoms may not be out of place here. Many are apt to suppose that there can be no better foundation than clay; and it is indeed true that some kinds of hard clay form a very satisfactory subsoil. But there are others of softer consistency, and permeated by sandy beds, which are extremely treacherous. Ardentallan Pier (Fig. 25) was erected in Loch Feochan, near Oban, in

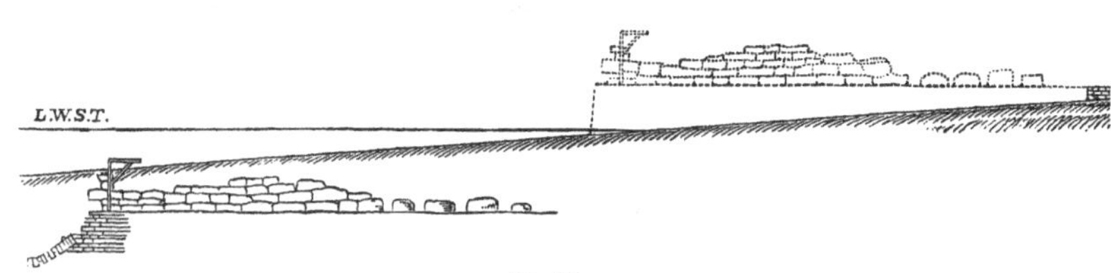

Fig. 25.

1838, and was used for the export of stone from a quarry of the same name immediately adjacent. The beach, at the place, sloped at the rate of about 1 in 6, its surface being coarse gravel mixed with clay and sand, which presented the appearance of a foundation of the firmest and most substantial character. Below low water the bottom dipped at the rate of about 1 in 12½. So satisfactory did the ground appear that the walls were founded on the surface without any excavation being made. The quay-wall forming the outer end of the pier was 120 feet in length of face, and was founded at the level of about a foot below low water. The hearting consisted of quarry rubbish and *terring*,* the length of quay, measured seawards from high-water mark, was 180 feet, and it was 6 feet thick and 12 feet high, with a face batter on that height of 9 inches. In 1838 there were upwards of 200 tons of

* Terring, from the Latin *terra*, earth, is the Scotch term for the soil removed in clearing the top of a quarry.

dressed stone deposited on the pier, and at that time it showed no symptoms of failure. In January 1844, however, while there were only 120 tons of material on the pier, the crane, which was on the outer end, suddenly began to move and shake; transverse openings also appeared, which soon got wider. The workmen, in much alarm, sent for the foreman from the quarry, who on his arrival saw, to his great consternation, the whole mass rapidly sinking and moving seawards, excepting about four feet at the land end, which remained firm. The quay at the outer end soon slid out at the bottom so as to incline at an angle of about 45°. In this way the whole pier, with the dressed stones on its top, rapidly slid outwards and downwards, and, within *two hours* from the first indication of motion, no part of the pier or even the crane could be seen; and on sounding it was found that the top of the quay was more than 23 feet below low water.

The *horizontal* movement was 150 feet in two hours, or at the rate of about 15 inches horizontally per minute, while in the same time the *vertical* movement was 34 feet, being 3.4 inches per minute. I ascertained these facts on visiting the ground after the accident, while the edges of the deep chasm which had been left were still fresh and unaltered. The clay forming the beach appeared to have resulted from the decomposition of the adjoining clay-slate rocks, and under the surface it was of a soft and almost semifluid consistency, and arranged in thin layers or films *parallel to the slope of the beach near high water, but dipping more quickly near low water.* The different layers of clay, which varied in thickness, from about one quarter to only about one-fiftieth of an inch, were separated from each other by thin films of sand. The layer of coarse gravel, which formed the hard surface on which the pier was founded, was only about 6 inches thick. The cause of failure seems to have been the slipping of some of the lower clay-beds, probably at one of the sandy seams, for I know of another case in which a clay-bank slipped forward on a similar thin sandy film. This accident shows the propriety of testing the subsoil by boring, or sinking trial pits, and the danger of resting satisfied merely with superficial appearances. It also proves the treacherous nature of stratified clay when associated with films of sand. Another somewhat similar case will be found under the next section relating to timber piers and quays.

Application of foregoing Principles to the Erection of a Building on a Sandbank.—This chapter may perhaps be best recapitulated by applying the principles it lays down to a case of known difficulty—viz. the erection of a building on a sandbank—such, for example, as the erection of a lighthouse on the Goodwin Sands. The reader may not perhaps have forgotten the disinterested attempts to erect "the light of all nations," as it was called, as well as the many chimerical projects for effecting the same object which have from time to time appeared; in all of which ingenuity was far more conspicuous than practical judgment.

The difficulties, as I have elsewhere shown, which have in such a work to be contended against, are principally the *reaction of the waves from the building upon the soft bottom;* and the mistake which was committed in all the designs for erecting a tower on this quicksand was in adopting the principle of *mechanical fixture* in preference to that of *dead weight.*

Without abandoning the known and fully-tried principles of marine engineering, the object, if it be one—for I express no opinion on that point—could be easily and certainly accomplished by simply depositing rubble stones in the sea, and allowing them to assume their own slope; and thus, after a year or two of labour, to form ultimately a small island of *pierres perdues*, similar to what was proposed in America in 1835 for Delaware Bay. In this operation there is neither novelty nor difficulty, nor any room for doubt as to success. This will appear by collecting together the requirements for such a work.

1. Pliancy of the mass of materials employed, so as to admit of this artificial island adapting itself without dismemberment to the changes of form consequent upon unequal settlement.

2. That the structure shall rise out of the sand at a low angle, so as to reduce to a minimum the reaction upon the quicksand.

3. That that part of the structure where reaction must to some extent occur shall be far distant from the place where the tower is to be erected.

4. That the connection of the structure with the bottom shall not depend upon mechanical fixtures.

5. That if any partial damage takes place during the erection, the ruins shall act as a check to farther encroachment.

6. That the waves shall not impinge with full force upon the tower itself, but shall be broken before they reach it.

CHAPTER VIII.

DESIGN OF GROUND-PLAN OF HARBOURS.

General Rules—Entrance Seaward of Works, and coincident with direction of heaviest Waves—Width of Entrance—Directions of Piers and Winds—Travelling Shingle—Good "Loose"—Curved and Straight Piers—Free and Confined Waves—Outer Basins—Ratio of Entrance to Area—Reduction of Waves by Lateral Deflection—Reduction under lee of Piers with free ends—Reduction in close Harbours—Formula for Reductive Power—Cellular Structure for reducing Waves—Stilling Basin—Situations for unprotected Quays—Booms—Small Harbours—Capacity of Commercial and of Fishing Harbours.

In laying out the general design or ground-plan of a harbour, the principal matters to be kept in view are the proper disposition of the lines of the piers, so as to insure safe and easy ingress and egress, and the inclosure and protection of a sheet of water of sufficient depth.

The positions in which the piers are to be placed depend on the nature and configuration of the shore and of the bottom. Before any step can be taken, the engineer must have before him numerous and accurate soundings, so as to give a correct representation of the bottom. The means of obtaining such data come strictly within the range of marine surveying, and I will not therefore enter at all upon the subject of these preliminary investigations, but shall leave the reader to consult those works which are specially devoted to this branch of surveying.*

* *Vide* D. Stevenson's "Marine Surveying:" Edinburgh, 1842.

After a correct plan, with soundings, has been obtained, the next step is to lay down contour lines of the different depths, which make the limits of the deep and shoal water at once obvious to the eye. The lines of the piers may then be sketched, so as, without sacrificing other conditions, to keep the works as much as possible on the shoal ground, while they at the same time inclose the greatest possible amount of deep water.

General Rules.—Many points requiring great care, for they affect vitally the ultimate success of the whole scheme, now present themselves for particular study. Among the most prominent of these—for we cannot take cognisance of all the peculiarities and the difficulties which may be found in every locality—are the following :—

1. *Entrance always Seaward of Works.*—The entrance should be fixed *seaward* of every other part of the works. If it be placed landwards, or even in line with a pier, we may cause an accumulated wave to be formed, which will pass into the harbour, or run across its mouth, rendering vessels unmanageable at the critical moment of "taking the port."

2. *Direction of the Entrance should, when possible, coincide with that of the Heaviest Waves.*—Unless where the internal area is small, and the sea very heavy, the direction of the entrance should be made to coincide with the direction of the heaviest waves, so that they may run along with, and guide vessels into the harbour. It is, no doubt, a simple and very efficacious mode of increasing the reductive power of a harbour to place the entrance at right angles to the line of movement of the swell. The old harbour of Pulteneytown furnishes a notable example of this disadvantageous arrangement. Such an expedient should indeed never be resorted to if it can possibly

be avoided; for when a vessel *takes* a port having an indirect entrance of this sort, she is liable to be struck by the waves on her broadside or quarter at the moment of turning in, and she thus runs a great risk of missing the harbour altogether and being stranded. If there be plenty of sea-room, this danger may be obviated by extending the outer pier sufficiently far seaward of the end of the other pier-head, so as to give it what we have called a free end, as in Fig. 26, and thus to allow a

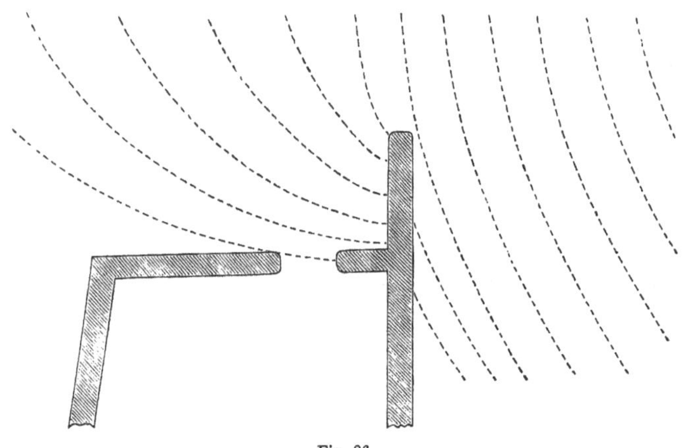

Fig. 26.

ship plenty of sea-room to shape an easy course. Even if, while rounding in, she should be struck by a wave so as to make lee-way, there is still sufficient time for her to recover herself under shelter of the outer breakwater before reaching the narrow entrance between the piers.

3. *Width of Entrance.*—The width of entrance channel adopted in France is, according to Minard, that of three vessels under sail, and this varies from about 100 to 330 feet. The following Table, compiled from various sources, gives the width of entrance of different ports :—

	Feet.		Feet.
Helvoetsluys (Minard) - -	108	Leith (Admiralty Pilot) -	240
Seaham (Admiralty chart) -	110	Yarmouth (J. Walker) -	250
Dover (J. Walker) - - -	120	Graveslines (Minard) - -	262
Newhaven (Admiralty Pilot) -	150	Sunderland -	315 and 368
Donaghadee (Tidal Harb. Rep.)	175	Do. South - - - -	230
New Shoreham - - -	176	Howth - - - -	315
Ramsgate - - - -	190	Calais - - - - -	328
Portland - - - -	400	Penzance - - - -	330
Ostende (Minard) - -	197	Aberdeen - - - -	378
Port Rush - - - -	210	Flushing (Minard) - -	426
Ayr (Tidal Harb. Rep.) -	215	Portsmouth - - -	700
Boulogne - - - -	230	Kingstown (Tidal Harb. Rep.)	960
Dunkirk (Minard) - -	236		

4. *Direction of Piers and Prevalent Winds.*—The piers should be laid out with reference to the direction of the prevalent wind. Minard says that at Bayonne, where they make an angle of 12° with the prevalent wind, they have been found to suit best the wants of the mariner.

5. *Danger of Travelling Shingle.*—If the beach consist of travelling detritus, the entrance must be so devised as not to form a trap for the passing shingle. This danger is most to be dreaded on the coasts where heavy waves strike obliquely upon the shore.

6. *Importance of a good "Loose" or Point of Departure.*—There must be a good "loose," so that vessels on leaving the harbour shall be able to shape their course free of rocks or a lee-shore, and Minard very properly remarks that the mouths of harbours should be designed rather to suit the entrance than the exit of vessels.* The following digest of 266 wrecks and casualties, occurring from different causes, is taken from the Harbour of Refuge Commission Report of 1859 :—

* Ouvrages Hydrauliques, p. 31.

Abandoned at sea from various causes		7
Burnt (fire caused by a cargo of lime getting wet) . .		1
Capsized		4
Damaged on piers or bars through insufficiency of tugs or tow-ropes		13
Explosions from coal gas generated under hatches . .		4
Foundered at sea and at anchors from various causes .		36
Leaky or dismasted and otherwise crippled (put back to repair)		22
Stranded or struck on rocks, bars, and piers, *in entering or leaving ports and harbours*, viz.—		
Blyth	1	
Hartlepool	12	
Hauxley	1	
Hird Sand	5	
Seaham	3	
Scarborough	3	
Shields	3	
Sunderland	27	
Whitby	3	
	—	58
Stranded or struck on rocks and sands from various causes .		119
Struck on sunken wreck		2
Total . .		266

This table shows that more than one-fifth of the whole casualties, occurring within the period mentioned, took place in entering or leaving port, and if to this we add those damaged on piers or bars through insufficiency of tow-ropes, we shall have upwards of one-fourth of the whole; or, in other words, the table shows that the mariner has often to encounter his greatest perils when he is nearest to his port of destination or departure.

7. *Piers of Horizontal Curvature preferable to Long Straight Piers.*—Preference should generally be given to a pier of

horizontal convex outline, or of a polygonal form, rather than to one long straight pier running at right angles to the worst waves. The principal objection to a straight pier does not, however, extend to cases where the heaviest waves strike upon it obliquely, and roll landwards along the sea-wall.

Free Waves and Confined or Gorged Waves.—Especial care, however, must in all cases be taken, that a pier nowhere presents to the sea a surface of *concave horizontal outline,* or, what is still worse, abrupt faces which form a re-entrant angle ; for the waves will then act with an almost explosive violence. For the breaking of a free wave is a very different thing from the breaking of a wave confined by a barrier of masonry. While the first may be compared to the harmless ignition of a loose heap of gunpowder, the other resembles the dangerous explosion produced by the discharge of a cannon.

It may in some cases be found better, in designing a *sea-wall* for protecting a sinuous coast, to carry the bulwark straight across ledges of rock which extend landwards, than to follow the line of the high-water margin. For although with the straight wall we have to encounter a greater depth of water and a heavier surf, still with the other we may have to oppose the waves with a wall which has a concave horizontal curvature, by which the force is concentrated and rendered far more destructive. Moreover, the straight wall may be considerably shorter than the curved.

8. *Outer Harbour or Stilling Basin.*—There should be sufficient distance landward of the mouth to allow a vessel, having full weigh on her, to shorten sail. For this Scanzin allows a distance of not less than 200 cables' lengths, and Minard recommends from 200 to 300 metres (1000 feet). Where the vessel requires to alter her course in order to reach the inner basin, no circle with a less radius than 200 yards

in smooth water should, I think, if possible, be adopted for the ordinary class of coasting steamers.

9. *The Relation of Width of Entrance to Area of Harbour.* —The internal area should bear such a relation to the width of entrance as to produce a sufficient degree of tranquillity, for which directions will be hereafter given (*vide* p. 146).

Reduction of Waves by Lateral Deflection.

Effects of Breakwaters in sheltering uninclosed Roadsteads.— No general remarks, that could prove of the slightest use in any particular case, can be made regarding some of the provisions which have been mentioned, but there are others of a more definite nature, and depending less on local peculiarities, which admit of further elucidation. Among these—and it is one of the most important—is the manner of estimating the reduction of the waves when they are deflected from their original direction and made to diverge into sheltered water. And I cannot but express regret that no attempt has been made, so far as I am aware, to obtain exact numerical results on this point, derived either from theory or experiment. I have been unable, indeed, to find that a single observation or experiment of any kind has been made upon the subject, and yet the whole benefits which are expected to result from the erection of our great national breakwaters depend entirely upon the reduction of the waves. When a wave encounters an obstacle such as a breakwater, if we suppose the portion which strikes it to be annihilated by the impact, or to be reflected seawards, the portion which is neither destroyed nor interrupted will pass onwards, and a part will spread laterally *behind* the breakwater. It is this reduction of height which it is so desirable to determine; and I hope that those who have opportunities of making observations on the effects of breakwaters in sheltering uninclosed roadsteads will be induced to take up the subject, which is undoubtedly one of great importance. My son, Mr. Robert Louis Stevenson, made some observations at angles of 45° and 90° at Pulteneytown breakwater, and these gave a coefficient of about .06 for the formula given in the next article, which, *for the class of waves observed*, would be in this case—

$$x = 1 - .06 \sqrt{\alpha^\circ}$$

where x represents the ratio of the reduced to the unreduced wave, and α the angle of deflection.

Reduction of Waves immediately under Lee of Piers with Free Ends.—When the waves are deflected by a pier with a free end, and run along its inner side, the reduction which they suffer will be due to the distance passed over, and to the angle of deviation produced by the pier. Though far from placing full reliance on so slender a stock of facts as I am in possession of, and which were partly the result of observations at sea, and of experiments made in a brewer's cooling tank about four or five inches deep, I may state that the amount of *reduction* in the height of an unbroken wave, after being deflected, was found to increase *directly as the distance traversed, and as the square root of the angle of deflection.* The following formula represents the results with the class of waves which were observed at different harbours, but it is given simply as approximate, and it is obvious that the coefficient must, as in the former case, be variable, depending on the kind of wave:—

$$x = 1 - .04 \sqrt{\alpha^\circ}$$

in which x represents the ratio of the reduced to the unreduced wave, and α the angle of deflection.

In the subjoined table are given the heights observed at the harbour of Latheronwheel :—

1853, Original height of Wave.	Distance passed over.	Angles.	Results by Formula.	Observed Reduced Heights.
4.5	16	51°	3.2	3.0
5.0	"	"	3.6	3.5
4.5	"	"	3.2	3.0
6.5	"	"	4.6	5.0
3.0	"	90°	1.9	2.0
3.5	"	"	2.2	2.5
4.0	"	"	2.5	2.5
5.0	"	"	3.1	3.0
4.5	32	140°	2.4	2.0
5.0	"	"	2.6	2.5
4.5	"	"	2.4	2.5
6.5	"	"	3.4	3.0

These observations were made at the outer kants of the pier of Latheronwheel, which is single, with a free end, and which acts on the waves in a different manner from a harbour which forms an *inclosed* area, to which I shall next refer.

Reduction in the Height of Waves after passing into Close Harbours.

The ultimate object of every harbour is to preserve the tranquillity of the inclosed area by lowering the height of the waves as they enter, and this property is variously possessed by harbours of different forms, and depends upon the relative widths of the entrance and the interior, the depth of water, the form of the entrance, and the relation between the direction of the entrance and that of the line of *maximum exposure.*

Formula for Reductive Power.—The only formula with which I am acquainted which gives the *reductive power* of a close harbour, is that which I published in the Edin. Phil. Journal for 1853. When the piers are high enough to screen the inner area from the wind, where the depth is tolerably uniform, the width of entrance not very great in comparison with the section of the wave, and when the quay-walls are vertical, or nearly so, and the distance not less than 50 feet from the mouth of the harbour to the place of observation, the following formula is applicable—

H = height of wave at entrance, in feet;
b = breadth of entrance, in feet;
B = breadth of harbour at place of observation, in feet; or more accurately, length of the arc with radius D;
D = distance from mouth of harbour to place of observation, in feet;
x = height of reduced or residual wave at place of observation, in feet.

$$x = H\frac{\sqrt{b}}{\sqrt{B}} - \frac{\left(H + H\frac{\sqrt{b}}{\sqrt{B}}\right)\sqrt[4]{D}.}{50}$$

If H be regarded as unity, x will come out a fraction which will represent what we may call the *reductive power* of the harbour at the point for which the calculation has been made; and by multiplying the height of any wave at the entrance by the fraction x, its reduced height is found.*

This formula is founded partly on the conclusions arrived at by Dr. Thomas Young, in 1807, in his theory of undulation, and partly on observations. These observations included the

* In the Encyclopædia Britannica the formula was stated in perhaps a better form, viz.—

$$x = H\left\{\sqrt{\frac{b}{B}} - \frac{1}{50}\left(1 + \sqrt{\frac{b}{B}}\right)\sqrt[4]{D}\right\}$$

but that given above is more convenient for calculation.

determination of the centres from which the waves expanded, and which appear to be situated not far from the middle of the entrance.

The method of applying the formula is to describe a circle on the ground-plan of the harbour from the point of union of the lines of the piers produced seawards, or (what is sufficiently near) from the middle point of the entrance. The radius adopted must be equal to the distance (D) between the centre of divergence and the place on the pier where the reductive power is wanted. The arc of a circle thus described must extend so far as to intersect the two side walls of the harbour, or in cases where one of the piers meets the shore at a shorter distance, the arc must be extended to the line of direction of the shorter pier produced landward of the high-water line. It is necessary to observe, however, that in such cases as the last named, where the shore intervenes, the formula is not applicable unless the beach slopes sufficiently to allow the waves to spend themselves freely. The distance B is then measured as a chord between the two points of intersection; or where the versed sine is large, B should be taken equal to the length of the arc. It is believed that this formula will be found to be of general application in all close harbours where the entrance is of a direct and simple nature.

In order to show the correspondence between the actual observations and the results of calculation by the formula, a table is subjoined, which contains averages of the observations at the different places mentioned. Those at Buckie are reduced from upwards of 2000 observations.

RESULTS of FORMULA for REDUCTIVE POWER of HARBOURS, compared with Observations.

Name of Harbour.	Calculated Reductive Power.	Observed Reductive Power.	Height of Wave in Interior of Harbour.		Remarks.
			A. By Calculation.	B. By Observation.	
Kingstown	.307	.374	2½ ft. to 3 ft.	3 ft. to 3¾ ft.	
Sunderland.					
1st. At stair on South Pier, at entrance to beaching ground . .	.440	.385	5.72 feet.	5.00 feet.	
2d. At South Dock . .	.1015	.1023	1.32 ,,	1.33 ,,	
Macduff.					
1st. 165 feet from inner entrance . . .	.44	.60	1.10 ,,	1.50 ,,	
2d. 340 feet from inner entrance . . .	.42	.532	1.05 ,,	1.33 ,,	
Fisherrow.					
1st. On west side . .	.478	.472	2.39 ,,	2.36 ,,	
2d. On east side . .	.430	.384	2.15 ,,	1.92 ,,	
Buckie.					
1st point of observation .	.649	.534	6.17 ,,	5.07 ,,	Mean of 244—*a*.
Do. do.	,,	.533	,, ,,	5.06 ,,	Mean of 117—*b*.
,, ,,	,,	.534	,, ,,	5.07 ,,	Mean of 127—*c*.
,, ,,	,,	.570	,, ,,	5.41 ,,	Results of storm, Nov. 1857—*d*.
2d point of observation .	.464	.323	4.41 ,,	3.07 ,,	*a*, as above.
,, ,,	,,	.338	,, ,,	3.21 ,,	*b*, ,,
,, ,,	,,	.309	,, ,,	2.94 ,,	*c*, ,,
,, ,,	,,	.394	,, ,,	3.74 ,,	*d*, ,,
3d point of observation .	.233	.207	2.20 ,,	1.97 ,,	*a*, ,,
,, ,,	,,	.229	,, ,,	2.18 ,,	*b*, ,,
,, ,,	,,	.186	,, ,,	1.77 ,,	*c*, ,,
,, ,,	,,	.342	,, ,,	3.25 ,,	*d*, ,,
4th point of observation	.186	.157	1.77 ,,	1.49 ,,	*a*, ,,
,, ,,	,,	.183	,, ,,	1.74 ,,	*b*, ,,
,, ,,	,,	.134	,, ,,	1.27 ,,	*c*, ,,
,, ,,	,,	.289	,, ,,	2.75 ,,	*d*, ,,
5th point of observation	.143	.136	1.36 ,,	1.29 ,,	*a*, ,,
,, ,,	,,	.162	,, ,,	1.54 ,,	*b*, ,,
,, ,,	,,	.113	,, ,,	1.07 ,,	*c*, ,,
,, ,,	,,	.158	,, ,,	1.50 ,,	*d*, ,,
6th point of observation	.119	.139	1.13 ,,	1.32 ,,	*a*, ,,
,, ,,	,,	.160	,, ,,	1.52 ,,	*b*, ,,
,, ,,	,,	.120	,, ,,	1.14 ,,	*c*, ,,
,, ,,	,,	.175	,, ,,	1.66 ,,	*d*, ,,
7th point of observation	.209	.222	1.99 ,,	2.11 ,,	*c*, ,,
,, ,,	,,	.289	,, ,,	2.75 ,,	*d*, ,,
8th point of observation	.274	.232	2.60 ,,	2.20 ,,	*c*, ,,
,, ,,	,,	.368	,, ,,	3.50 ,,	*d*, ,,
9th point of observation	.427	.339	4.06 ,,	3.22 ,,	*c*, ,,
,, ,,	,,	.491	,, ,,	4.66 ,,	*d*, ,,

Mean of *all* the *calculated* heights in column **A** . . . . 2.64 feet.
Mean of *all* the *observed* heights in column **B** . . . . 2.62 ,,
Mean of the *observed* heights, taking under Buckie only those marked *a*, and those marked *c* where *a* is wanting (at points 7, 8, and 9) . . 2.47 ,,
Mean of *observed* heights in **B**, taking only those marked *c*, under Buckie 2.41 ,,

Side Channels for reducing Waves.—At the harbour of West Hartlepool an ingenious and novel device for reducing the height of the waves has been carried out by Mr. R. Ward Jackson and Mr. Casebourne. Interior expansions have been made to communicate with narrow canals running landwards, and which ultimately join the sea outside of the harbour. The portion of the wave which has been detached by spreading into the lateral channels is thus conducted entirely out of the harbour into the open sea. At Mullaghmore, County Sligo, where the *run* was troublesome in stormy weather, the basin has, I am informed, been much smoothed by an opening made at the upper end of the harbour, through which the swell passes out to the beach, instead of being reflected by the inner walls.

Cellular Structure for reducing Waves; "clair voie."—There are many places so narrow and confined by rocks as not to admit of the formation of lateral expansions either of the usual kind or of that adopted at Hartlepool. In such situations some reduction might, I think, be effected by converting the upper portion of the quay-walls at and near the entrance into a series of chambers, separated from each other by *vertical* diaphragms, so as to smooth the water by forming numerous *stops.* This cellular structure, in some respects similar to what is called in France "*clair voie*" (which is found to answer best with $\frac{1}{3}$ of solid to $\frac{2}{3}$ of void), might be cheaply constructed of vertical partitions of timber. The action of the cells consists in abstracting from the upper part of the wave a small portion of water, and retaining it momentarily until the crest has passed the mouth of the cell, when the water so retained falls again into the harbour on the *back* of the wave from which it had been abstracted. If the entrance-passage were of some length, even although the openings were of small width, a considerable reduction would be produced by this repeated process of separation and detention. The tranquillising tendency of such *stops* may, to some extent, be estimated by noticing the action on a passing wave of vertical fenders on the face of a pier. On a somewhat similar principle the waves are sometimes reduced by making a portion of the outer piers of *open* timber work, which allows the wave to burst through it.

Stilling Basin.—As before mentioned, it is essential when the exposure is great that there be either a considerable internal area, or else a separate basin opposite the entrance to the inner basin, for the waves to destroy or *spend* themselves. Such a basin should, if possible, inclose a portion of the original shore for the waves to break upon, and when circumstances preclude this, there should be a flat talus of at least 3 or 4 to 1, as recommended by the late Mr. Bremner of Wick. Mr. Scott Russell has found that talus walls of 1 to 1, or steeper, will not allow the waves to break fully, but will reflect them in such a manner as might in some cases make the entrance difficult or even dangerous of access, and the berthage within unsafe; and I can corroborate this from personal observation. Instances are not wanting of harbours being materially injured by the erection of a vertical wall constructed across a sloping beach on which the waves were formerly allowed to expend their force. The experiments made by the committee of the British Association exemplify this. Small waves from one quarter to half an inch high, which were generated in a box 20 feet long, could be traced after being reflected 60 times, and after having passed over a space of 1200 feet.

Situations where unprotected Quays may be used.—The following cases, in which traffic has been successfully carried on at quays unprotected by covering piers, may be found useful as a guide:—

At Scrabster the quay is at right angles to a fetch of 6 miles.
At Invergordon there is a fetch of 5 miles.

At Burntisland	,,	6	,,
At Kilcreggan	,,	4	,,
Londonderry	,,	1½	,,
Greenock	,,	6½	,,
Albert Quay, Greenock	,,	7	,,

The two last are, however, somewhat sheltered by the Greenock bank.

So far as my experience goes, I think that 5 miles is not far from the limits that should be observed, which by the ordinary formula (p. 23), $h = 1.5 \sqrt{D}$, gives a wave of 3.3 feet; but the formula for distances so short as these would give waves about a foot higher, and in heavy storms perhaps waves not much less than 5 feet may exist, but on such occasions vessels could not safely use the quay.

Open Timber Wharves useful for preventing Waves from being reflected.—Waves, when not of large size, can often be either *destroyed* or prevented from being reflected by adopting a sloping face of stones, with open timber quay in front.

Booms for excluding Waves.—In order to tranquillise harbours of small reductive power, logs of timber, called booms, having their ends secured by projecting into grooves cut in the masonry on each side, are placed across the entrance of the inner basin or dock. From 10 to 20 logs are usually dropped into those grooves, or as many more as will insure close contact of the lowest log with a sill-piece placed in the bottom of the harbour. They are also warped down or fixed with an iron hasp at the coping course, without which precaution the swell is found to enter the harbour from underneath. By this contrivance, which forms a temporary wall, the waves are checked, and completely prevented from spreading into the interior basin. The longest booms I have seen are at Banff, where the span is 43 feet; and in some places, as at Hartlepool and Seaham, in Durhamshire, they are taken out and in by steam-power.

Though booms are perfectly successful in their tranquillising effect (provided they are kept in contact with the sill-piece at the bottom), yet they are not suited for harbours where there is much traffic, as the "*shipping*" and "*unshipping*" of so many logs of timber involve a delay which might be attended with serious consequences, though this risk may be materially reduced by the employment of hydraulic power. Hollow booms, constructed of boiler-plate on the tubular bridge principle, might be found suitable for large spans; and if their buoyancy were destroyed by making them pervious to water, they would not require to be warped down, as is necessary when logs are used.

Capacity of Harbours for Commerce.—The capacity of commercial harbours for trade varies so much with the exposure, and size of vessels, that it is difficult to approximate to the truth. At Ramsgate, for instance, there were found to be about 6 vessels to an acre in the outer harbour, while there were about 14 in the inner and better protected basin.

Capacity of Fishing Harbours.—In the Scotch fishing harbours the number of boats used to be reckoned at from 85 to 115 per acre; but of late their size has been much increased, and probably not more than from 80 to 90 could be accommodated. The Cornish boats at New Lynn, according to the Channel Pilot, vary from 60 to 80 per acre.

Small Harbours in some respects more difficult to design than Larger Ones.—I cannot leave this part of the subject without observing that in some particulars the difficulties of design are inversely proportional to the extent of the works. Indeed, if the piers inclose a very large area, some of the elements of difficulty nearly altogether disappear. Little attention need then be given to those questions which are so troublesome in small basins regarding reductive power, want of spend, and recoil of waves; and comparatively little as to the width of the entrance.

CHAPTER IX.

DOCKS, TIDE-BASINS, LOCKS, GRAVING-DOCKS, SLIPS, ETC.

Advantage of Docks—Entrances to Docks and Slips—Outer or Tide Basins—Proportions—Locks—Dock of Maximum Capacity—Available Capacity of Quays—Shelter and Capacity of Docks: Amount of work done at—Graving Docks—Iron Floating Dock—Hydraulic Lift—Patent Slips—Gridirons—Hydraulic and Screw Docks—Relative Advantages of Slips and Graving Docks—Shipbuilding Stances—Dock-gates.

ON applying our formula (p. 147) to any design for a harbour, we shall soon find whether it be possible to secure a sufficient amount of sheltered space within the piers. If the calculation shows that this cannot be done, the best course will be to divide the inclosed area into an outer and inner harbour; and if the reductive power even of the inner basin be still found too small, a dock or tide basin with gates may then be resorted to. But in considering the eligibility of docks we must also remember that the principal advantages which they confer are the great conveniences for trade, which warrant their adoption in places where no protection from the waves is required.

Advantages of Docks.—The peculiar advantages afforded by docks are the following:—Vessels can be accommodated in the smallest possible space, and are enabled to lie constantly afloat; whereas in tidal harbours where they *take the ground* they are apt to be strained, or to have their floors broken. But there are other sources of mischief than this, for often, when the tide is ebbing, vessels, unless watched, fall against each other. In two instances, one of which gave rise to subsequent legal proceedings, where the bottom, which was muddy, had a considerable declivity, a ship, which had taken the ground on the beach near low-water mark, was actually run down and damaged by another stranded vessel, the warps of which suddenly snapped and freed her from the moorings at the quay, thus causing a collision between vessels, both of which were at the time high and dry. Then there is the chafing of the vessels' sides against the quays, and the breaking of warps during stormy weather, or during land floods where there is a river. It is said that at Sunderland damage to the extent of £40,000 was occasioned in one day by large quantities of ice that came down the Wear. When a vessel is in dock, she can be easily and at all times moved from place to place, while the operation of discharging and loading can go regularly on during any time of tide. Her level, too, is never much affected, so that the cargo does not require to be hoisted so high as would otherwise be necessary.

The relative eligibility of docks or tide basins depends much on the rise of the tide. In such rivers as the Clyde and the Foyle, where the lift does not exceed about 9 or 10 feet, docks are less needed than at the Mersey, Bristol Channel, and similar places, where the rise is from 30 to 50 or even 60 feet.

Entrances to Docks and Slips.—Mr. Redman, who has devoted much attention to the subject of the entrances to docks placed in a tideway, says "the practice in the port of London is to dock a ship upon the flood just before high water, and to undock her at about the same period of tide. . . . The directions apparently the most desirable are an angle of about 45°, pointing up the stream, for graving docks, and an angle of about 60° for wet docks." * Each case must, however, be judged on its own merits; as, for example, where there is a liability to the deposit of silt, it may be better that the entrance should point down stream.

Outer or Tide Basins.—The lock-gates through which vessels enter a dock are thrown open, as at Liverpool for example,† for about two hours at high water, so that vessels can pass freely out and in. But as the tide falls they must be shut, otherwise the level of the water in the dock would be too much lowered. After that time of tide vessels can only enter or leave by being locked up to the dock or locked down to the sea. Outer or half-tide basins, first introduced, it is believed, by the late Mr. Jesse Hartley, are formed between the lock and the sea, and admit of a large additional traffic being accommodated. The entrance to such basins is provided with sea gates, which are kept open till half-tide, or such other time of ebb as is found most suitable to the situation. Long after the lock-gates which form the entrance to the docks are closed, inward-bound vessels can run into the outer basin, and be afterwards passed into the docks; and in like manner outward-bound vessels can be passed down from the dock long after high water.

The Proportions of Outer Basins to Dock Areas.—The relative proportions of area of the outer basins to docks vary at different ports, according to the amount of traffic to be provided for. The following Table shows the proportions at one or two harbours:—

* Min. Civ. Eng. vol. xviii. p. 495. † Artizan, vol. iii. p. 221.

PROPORTIONS of DOCK AREAS to AREAS of OUTER BASINS.

		Acres.	Ratio.
Bute Docks (East) Cardiff -	Basin	2.15	1 : 20
	Dock	42.33	
Do. (West) „ - -	Basin	1·244	1 : 14
	Dock	16.86	
Penarth - - - -	Basin	2.68	1 : 65
	Dock	17.11	
Tyne - - - -	Basin	9.5	1 : 5.25
	Dock	50.0	
Liverpool (General) - -	Basin	13.0	1 : 8.23
	Docks	107·0	

Locks.—It is a curious fact, illustrative of the strength of prejudice, that, when Whitworth proposed to form a lock at Leith, in 1786, it met with strong opposition, on the ground of its being dangerous to shipping. Such an objection, it is almost needless to add, is never now heard of.

The dimensions of locks depend, of course, on the class of shipping that has to be provided for. In France the amount of *side-play* that is allowed for merchant vessels is generally .65 foot, and 1 foot for vessels of the first class. Minard allows from about 4 inches to about 10 inches between the sill and the vessel's keel. At Flushing he provided about $\frac{6}{10}$ foot for vessels drawing 24 feet; and he has seen ships of war pass through with only 6 inches under their keel. The following table, taken, by permission, principally from the late Mr. Beardmore's Tables, shows the dimensions of some of the locks and docks at different ports :—

TABLE.

	Length of Lock.		Breadth.		Depth over Sill at High-water Springs	
	Ft.	In.	Ft.	In.	Ft.	In.
Leith (East) - - -	160		36		18	5
Dundee (Victoria) - -	230		60		21	
Aberdeen (Victoria) - -	250		60		21	
Dublin (Large Dock) - -	180		36		18	
Cork - - - - -	180		45		18	6
Bristol (Cumberland) - -	260		54&67		30	
Plymouth - - - -	250		55		24	
Newport - - - - -	225		61		25	
Cardiff - - - - -	152		36		19	
Swansea - - - - -	165		56		21	6
Ipswich - - - - -	150		45		16	6
Hull (Humber) - - -	158	6	42		24	
Great Grimsby - - -	300		70		26	
Goole (Railway) - - -	266		58		19	
„ (Barge Dock) - -	72	6	19	6	9	
Middlesborough - - -	132		30		19	
London (Victoria) - -	326	6	80		28	

Lockage.—The late Dr. Rankine, in his *Manual of Civil Engineering*, gives the following table for the expenditure of water due to the passage of vessels through the lock :—

Let L denote a lockful of water, that is the volume contained in the lock chamber between the upper and lower water levels.

B the volume displaced by a ship.

Then the quantities of water discharged from the dock are shown in the table. The sign – prefixed to a quantity of water denotes that it is displaced from the lock into the dock.

	Lock found	Water discharged.	Lock left
A descending ship . .	Empty . .	L − B	Empty.
do. do. . .	Full . .	− B	Do.
An ascending ship . .	Empty or full	L + B	Full.
2 n ships descending and ascending alternately	Descending full Ascending empty	n L	Descending empty. Ascending full.
Train of n ships descending	Empty . .	n L − n B	Empty.
Do. do. .	Full . .	(n − 1) L − n B	Empty.
Train of n ships ascending	Empty or full	n L + n B	Full.
Two trains of n ships—the first descending, the second ascending	Full . .	(2 n − 1) L	Full.

From these calculations it appears that ships ascending and descending alternately cause less expenditure of water than equal numbers of ships in train.*

TABLES showing DIMENSIONS of different DOCKS.

	Length.	Breadth.
Sunderland - - - - -	645 feet.	147 feet.
Leith (East) - - - - -	250 „	100 „
„ (Victoria) - - - -	233 „	200 „
Aberdeen - - - - -	950 „	175 „
Dublin (large) - - - -	217 „	100 „
Galway - - - - -	239 „	193 „
Limerick - - - - -	270 „	130 „
Bristol (Cumberland) - -	245 „	90 „
Plymouth - - - - -	420 „	150 „
Newport - - - - -	270 „	73 „
Swansea - - - - -	760 „	80 „
Great Grimsby - - - -	600 „	167 „
Hull (Victoria) - - - -	480 „	126 „
Goole (Barge) - - - -	290 „	50 „
„ (Ship) - - - -	234 „	67 „
„ (Steamer) - - - -	120 „	164 „
Middlesborough - - - -	400 „	130 „

* Manual of Civil Engineering, by W. J. M. Rankine; London, 1862, p. 751.

Dock of Maximum Accommodation for Trade.—Fig. 27* is an attempt to represent the form of maximum capacity for a dock, and although in most cases, owing to local circumstances, it may not be possible of adoption, yet it may be sometimes at least approximated to. Internal

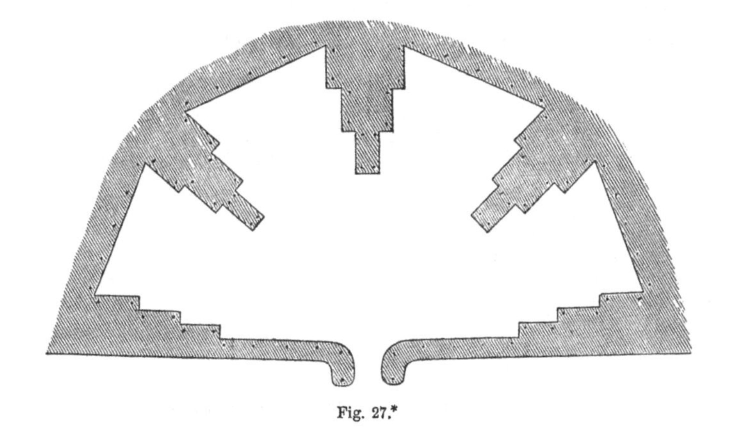

Fig. 27.*

jetties with a broken line of quay are in common use, but the radiating arrangement has not, so far as I know, been adopted. The object is to secure as much room as possible at and near the entrance.

Available Capacity of Quays.—The amount of work that can be done per yard-length of quay varies with the different facilities which are afforded for traffic, as, for example, whether common cranes or steam or hydraulic cranes are used, or whether there are ordinary carting or tram-ways connected with the shipping berths, or whether the depth of water in front is small or great, or whether the quay is exposed to the rise and fall of the tides as in a river, or whether the ships are always water-borne as in a dock. The available free space of ground behind a quay forms also an important element as regards traffic. Wherever possible, it is desirable to have at least 100 feet of breadth behind the quay. At Glasgow there is generally allowed 115 feet of space. Vessels of 150 tons require about 100 feet of quay.

The following Table, given in Mr. Ure's report on the extension of Glasgow harbour in 1854, gives the amount of work done per annum, per yard of quay and per acre, at different ports:—

	Tonnage per Acre of Water Space.	Tonnage per Lineal Yard of Quay.	Remarks.
Glasgow, including steamers	30,670	440	Fully worked.
,, excluding do.	20,361	293	
Southampton tidal harbour	20,000	350	Could do more.
Liverpool Docks . .	21,200	185	Fully worked.
St. Katharine's Docks .	20,500	142	Do. do.
Hull Docks . . .	19,000	143	During nine months.
To this may be added the old Bute Docks, which did 788,960 registered tons, or—	39,448	255	
Tyne Docks . . .	40,000		

TABLE of PROPORTIONS of QUAY per Acre of water space.

	Lineal Feet.		Lineal Feet.
East India - - -	216	Nelson (Liverpool) -	300
Table Bay - - -	240	Salisburgh ,, -	393
Southampton - -	180	Collingwood ,, -	330
Birkenhead - - -	474	Stanley ,, -	321
Belfast (proposed) -	216	Clarence ,, -	456
Huskisson, Liverpool -	225	Trafalgar ,, -	381
Sandon ,, -	260	Victoria ,, -	411
Wellington ,, -	306	Princes ,, -	405
Bramley ,, -	282		

The above is from a statement of Mr. Brunlees, and from other sources. Sir John Coode states that the proportion is generally from 200 to 250 feet per acre.

Docks should be sheltered from Wind.—After the engineer

has succeeded in designing a dock which is sufficient, in so far as the sea is concerned, it may after all not prove safe and convenient, if it be exposed fully to the force of the wind, which, acting on the rigging and hulls of the shipping, produces a grinding action of the vessels against the quays. At Sunderland south dock, and at the docks in the Tyne, the gales of October 1863 occasioned very considerable damage, from vessels breaking adrift from their moorings, and coming into collision with other vessels.

Capacity of Basins and Docks for Trade.—The number of vessels that can be accommodated in each acre of a basin may be termed its *available capacity.* This must obviously vary with the size of the craft which frequent the port, and with the ratio of sheltered to unsheltered acreage, or, in other words, with the exposure and the reductive power. It will therefore be highest for a dock with gates, less for an outer basin into which the waves have access, and least of all for an anchorage-breakwater or roadstead. But it also depends on the form of the basin, and on its area in relation to the class of shipping to be accommodated. It is hoped that a tolerably good approximation to the capacity of a dock will be found by the formula

$$n = \frac{1000}{t} + a$$

where n represents the number of vessels per acre, and t their average tonnage. a is a coefficient which may perhaps be taken at from 4 to 5.

Tonnage.	Number of Vessels per Acre, by Formula.	Tonnage.	Number of Vessels per Acre, by Formula.
100 - -	14.0	350 - -	6.9
150 - -	10.6	400 - -	6.5
200 - -	9.0	450 - -	6.2
250 - -	8.0	500 - -	6.0
300 - -	7.3		

Commercial Value of Docks of different Depths.—The capacity depends not on the area only, but on the depth as well. The proportions of depths to tonnage are referred to in Chapter XI.

Sir J. Hawkshaw has introduced the plan of erecting a series of separate jetties on the sides of docks, instead of constructing continuous quay-walls. A large traffic can in this way be accommodated at a very considerable saving of cost. Such jetties, being formed of timber framing, render the expense of stone foundations resting on pile-work unnecessary, while they admit of the construction of continuous quay-walls of masonry, when a future expansion of the trade of the port justifies that additional expenditure.

Graving Docks, Patent Slips, Gridirons, etc.

Although graving docks are generally unremunerative, yet they largely tend to raise the character of a port; and hence almost all harbours of any importance have either graving docks, Morton's patent slips, or gridirons.

Graving Docks are basins, fitted with gates, from which the tide-water which floats the vessel into the dock is pumped out, so as to let the carpenters get access to the ship's bottom. The sides of graving docks consist of a series of steps of masonry, called *altars*, against which small timber props, generally of Gulf of Bothnia timber, are placed, for supporting the vessel's sides as she ceases to be water-borne. Her keel is supported on blocks, generally of hard wood, but of late years they have in some places been made of cast iron. The sides, in order to save pumping, are in some places made of a curved form, so as to suit the shape of the vessel's sides. The advantages are, however, more than counterbalanced by the undue contraction of the space allotted for the carpenters, who are unable to move about easily on the sloping surface of the masonry.

Of all the different kinds of masonry which enter into the construction of marine works, there is none which requires greater accuracy of workmanship, or more careful circumspection, than the graving dock. Leakage in a wet dock, provided it does not originate at a place where it is liable to increase through time, and is of no greater extent than to depress the surface of the water a few inches, cannot be regarded as a serious evil. But in a graving dock, where the requirements are different, there should be no leakage. A very little water, accumulating on the platform of a dry dock, interferes to a serious extent with the comfort and convenience of the carpenters. Although it may occasion considerable additional expense, there ought, in all cases where the soil is full of springs, to be an ample underground storage provided, by a system of drains, for receiving the leakage, which can then be pumped out periodically, as required, without ever allowing the water to rise above the platform.

Mr. G. B. Rennie's Iron Floating Dock.—These docks are stated to be the first of the kind that have been made of iron. They consist of floating caissons for holding the vessel to be repaired. They are sunk by allowing them to fill with water, and are raised by pumping. The caissons are made with water-tight compartments, and they are carried up as high as the vessel's bulwarks, excepting that through which the vessel enters and leaves. Among several advantages tha have been claimed for this kind of dock may be mentioned—its independence on the rise and fall of the tide, the power of applying breast shores as in an ordinary graving dock, and the stiffness produced by the side walls. As the upper parts of the side walls or altars are always full of air, this dock may be used in deep water, and is therefore independent of the nature of the bottom.

Mr. Edwin Clark's Hydraulic Lift, which is situated at the Victoria Docks, London, consists of a pontoon which is filled with water and sunk between two rows of iron columns. After a vessel has been floated and steadied upon the pontoon, the whole is raised by twelve hydraulic force-pumps of 2 feet diameter, acting on the pontoon by means of chains. After the pontoon has been brought above the tide-level, the water is allowed to escape, when there is sufficient floating power to admit of the whole being removed to any place where the repairs can be conveniently made. The pontoons which can accommodate vessels of a length of 350 feet are about 320 feet long and 59½ feet broad. "The power of the hydraulic lift is 6400 tons. The largest pontoon will carry a dead load of 3200 tons in addition to its own weight."

Patent Slips are the contrivance of the late Mr. Thomas Morton of Leith, and consist of a carriage or cradle working on an inclined railway, falling generally at the rate of about 1 in 17, and extending from several feet above high water to several feet below the level of low water, and a truck or carriage which moves on it. When this carriage is let down under the water, the vessel is floated above the place, and the carriage is drawn up till it catches the vessel forwards. When the ship is placed truly above the line of the carriage, a powerful crab-purchase at the top of the slip, which is generally worked by steam, is set in motion, and raises the truck and ship out of the water.

Mr. Redman recommends that slips which are placed in a tideway should be laid out at a right angle to the axis of the stream.

The Gridiron is a simple framework of timber placed at a level sufficient to admit of vessels being floated above it during the flood-tide, and grounded upon it during the ebb, and when thus left high and dry the vessel's bottom can be examined to ascertain if it be necessary to take her into the graving dock, and trifling repairs can also be made. The gridirons at Liverpool vary from 25 feet to 36 feet 3 inches in breadth, and from 228 feet 3 inches to 313½ feet in length.*

The Hydraulic and Screw Docks used in America are chambers into which vessels are floated during the flood-tide, above a cradle which is drawn up above the high-water level, either by means of Bramah's press worked by a steam-engine, or by a powerful apparatus of screws.

Relative Advantages of Slips and Graving Docks.—The advantages which the patent slip possesses over the graving dock may be said to be—*First,* The cost of its construction is less. *Second,* When the fall of tide is languid, a vessel can generally be more quickly laid dry. *Third,* When so laid dry she can be more easily examined, and from the duration of the daylight being greater than in a deep graving dock, the hours in which work can be done are in this country extended during winter about forty minutes per day. *Fourth,* There is more perfect ventilation, by which the vessel's sides are sooner dried, which is of some moment with an iron ship. *Fifth,* A vessel can be hauled up long before high water, and the repairs can be at once begun; whereas with a dock the pumping occasions considerable delay. *Sixth,* While the upper part of the slip is occupied, an additional vessel may be taken up for a shorter time than its predecessor, without interrupting the workmen.

The advantages afforded by a graving dock, on the other

* Historical and Descriptive Sketch of the Mersey Docks and Harbour, by J. J. Rinckel. Published in the *Artizan* for 1864.

hand, are—*First*, That, although its construction is more costly, it is nevertheless, if properly built, unquestionably a more durable structure—the rails, rollers, carriages, and chains, connected with the slip, being liable to derangement, which entails occasional repair. *Second*, The management of the graving dock is simple, and involves comparatively little superintendence; whereas that of a slip is intricate, and requires more than mere nautical skill. *Third*, The working of a graving dock is equally simple for large or small vessels, while it is undeniable that the raising of a large vessel on a slip is a delicate operation, and should be attempted only under the direction of persons thoroughly versed in such matters, and having ample mechanical resources at command. *Fourth*, The graving dock possesses the advantage, which is sometimes important, of affording the means of more easily filling a vessel with water, so as to detect leaks which may not be discoverable by other means. For this purpose a nozzle to receive a flexible tube should be fixed into the dock-gates. *Fifth*, Where double gates are provided, the water contained in the dock affords a certain limited power of scouring the forebay and entrance, an advantage which is of course not possessed by a slip. *Sixth*, In any rapid land current, or strong tideway, it is a much easier process to dock a vessel than to land her safely on the cradle of a slip—an operation which, when incautiously gone about, has been in some cases attended with serious consequences even in sheltered situations. *Seventh*, The graving dock need not interfere with the set of the currents, whereas a slip which projects a long way seaward of low water may deflect them and produce shoals in the channel. *Eighth*, Mr. Mallet has remarked that the strains on ships' timbers are more direct than when she is on a slip, especially when she is leaving the cradle. The late Mr. J. M. Balfour suggested, in order to meet this objection, that the cradle for a slip might be made of a wedge shape, so that its upper surface shall be parallel with the horizon, or that the back end should even be tilted slightly, so as to give a *bite* on the vessel and prevent her from slipping.

The relative advantages of the other contrivances for the repair of ships already described may be judged of by comparing them with each other, in a similar manner to that which has been done with the graving dock and slip.

Shipbuilding Stances.—A frontage of from 60 to 70 feet is about the average width which may be allowed for each shipbuilding stance.

Dock-Gates.—Mr. P. W. Barlow has given the following formulæ for the strength of dock-gates :—

Formula for Straight Gates.

φ = horizontal angle (or "sally") between pointing sill and line joining heel-posts of the two leaves.

W = pressure on the length of the gate with any head and for a given depth of the gate.

S = whole transverse strain at angle φ.

$$S = \tfrac{1}{2} W \sec. \varphi + \frac{1}{20} \text{cosin } \varphi.$$

From this Mr. Barlow has deduced that the salient angle, where the strain is the minimum, is 24° 54′, but as the length of the gate increases with the secant, the strength will not at this angle be the greatest with a given section of timber.

The *sally* or angle which gives the greatest strength, with a given section of timber, is stated by him as 19° 25′.

Formula for Curved Gates.

When θ is the salient angle, or camber of the beam, formed by a chord line drawn from the heel to the mitre-post, with the tangent to the curve of the gate—

$$S = \tfrac{1}{2} W \left\{ 1 - \frac{\sin. \theta}{\sin. (2 \varphi - \theta)} \right\}$$

There is great difference of opinion among engineers as to the strain to which dock gates are subjected, and the reader is referred for further information to the 18th and 31st volumes of the Minutes of Proceedings of the Institution of Civil Engineers. In these discussions Mr. Brown pointed out an error in Mr. P. W. Barlow's paper, which stated that the line of pressure at the mitre-posts would always be a tangent to the curve of the separate gates, whereas that line must always be at right angles to the centre line of the lock, and could only be a tangent to the curve when the two gates formed a segment of a circle, or as Mr. Bramwell says, at all events, that their junction at the mitre-posts formed, at that point, part of a continuous curve. Mr. Brown gives elaborate formulæ suited to meet a yielding or deflection of the structure, which he alleges must always take place. Mr. Bramwell states, and we think justly, that when the gates form when closed a segment of a circle they cannot be subjected to transverse strain, and that the whole of the gates would be subjected simply to compression. Mr. R. P. Brereton very properly suggests that when the gates are of malleable iron the boiler-plate should never be less than $\frac{1}{4}$ of an inch thick, whatever the formula may indicate.

Where l represents the length of one half of a straight or cambered malleable iron gate, w the distributed pressure over the length of the leaf taken on a given element of the gate, bounded by two horizontal planes one foot apart, t the thickness of framework of gate or distance between the two skins, s the transverse strain in middle of gate, θ half of the mitreing angle—*i.e.* the angle formed by meeting of gates—all the dimensions being in feet, and weight in tons—

$$s = \frac{\frac{1}{2}\, w\, l}{4\, t}$$

$\frac{s}{4}$ = sectional area of metal on compressed side in inches.

$\frac{s}{5}$ do. do. on extended side do.

$\frac{1}{2}\, w \tan \theta$ = compressive strain produced by other leaf of gate.

Mr. Kingsbury adds half of this compressive strain to the strain representing the compression due to the transverse pressure, and deducts it from the same amount for the extension; and these results being divided as before by 4 and 5 respectively, being the allowances per square inch of metal for compression and extension, give the areas for each. He, however, expresses a doubt whether the whole of the compressive strain from the other gate may not perhaps come upon the compressed section. The sections close to the posts he takes as requiring to resist the compression due to transmitted pressure only, or $= \frac{\frac{1}{2}\, w \tan \theta}{4}$.

For Cylindrical Gates.—Mr. Kingsbury gives :—When p = the pressure per unit of surface, and r the radius of curvature—

$$\frac{p\, r}{4} = \text{sectional area of metal in square inches.}$$

Gates placed in exposed situations.—It must be remembered, in designing gates which are exposed to the waves, that they should be made stronger than the formulæ require. At Seaham the gates are 35 feet in width; and on the supposition of the maximum waves outside being 13½ feet in height, as at the port of Sunderland, which is only a few miles farther north, the formula for reduction (p. 147) indicates 4 feet as the height at the sea gates. But the dock is not available at such times, for whenever the waves exceed about 2 feet, it is found that the gates cannot be properly worked. At the north dock of Sunderland Mr. Meik informs me that the gates, which are 51 feet wide, cannot be worked when the waves are 3 feet high; while the greatest height of waves at the South Dock gates, which are 60 feet wide, is 8 feet, and they cannot be worked when the waves are 2 feet 6 inches high.

The gates for graving or dry docks sustain a greater pressure with the same rise of tides than those of wet docks, and should therefore be made correspondingly stronger.

Examples of gate-construction, as carried out at different works, will be found in the Plates XIV. XV. and XVI.

Dock Walls.—Mr. Giles recommends that dock walls should always be made of sufficient strength to resist a pressure of water equal to their height. Minard assigns four-tenths of the height for the thickness, and Rankine takes the ordinary thickness at from one-third to one-half of the height. Some examples will be found in Plate XVII.

CHAPTER X.

MATERIALS, KINDS OF MASONRY, IMPLEMENTS, ETC.

TIMBER: Early Use—Marine Insects—Experiments at Bell Rock—Failure of Greenheart-protection—Pile-work—Screw-Piles—Destruction of Stone.
IRON: Decay of—Mallett's Experiments—Dressing and Mode of assembling Masonry—Settling of Rubble—Size of Materials—Beton and Concrete—Lifting and Setting Blocks—Monolithic Structure-staging.

TIMBER.

Early Use of Timber.—The employment of timber in harbour works is of great antiquity. It seems first to have been used in forming boxes which were filled with stones, and at a more recent period in the formation of open frameworks through which the current could pass freely. Vitruvius mentions moles, consisting of timber filled with stones and cement, as having been used by the harbour-builders in his days. These structures, says he, "which are built in the water, are thus executed. The sand of the country, which extends from Cumæ to the promontory of Minerva, is procured and mixed with lime in the proportion of two to one; then in the intended place, fences of oaken piles, bound with chains, are fixed in the water, and firmly united. When this is done, the ground at the bottom of the water between these fences is by means of transtilli (?) cleaned and levelled, and rubble stones, with mortar mixed in the manner before written, is thrown in, till the space between the fences is quite full." * It was probably to some such work that Horace refers in Ode 1st, lib. iii.—

> " Contracta pisces aequora sentiunt,
> Jactis in altum molibus ; huc frequens,
> Caementa dimittit redemptor."

A harbour contract, in all probability the earliest which has been preserved, appears in the "*Registrum Nigrum de Aberbrothoc*"— a collection of ancient documents printed by the Bannatyne Club, and edited by Mr. Cosmo Innes, Professor of Antiquities in the University of Edinburgh. The contract, which is between the Abbots and Burgesses of Aberbrothoc (Arbroath), bears date 1394, and provides in the following terms for the erection of a haven by the Abbots, while the Burgesses undertake to find the materials and to clear the ground :—

" Tandem . . . . in hunc modum restat concordatum videlicet quod Abbas et conuentus supradicti loco peritorum indicio eminenciori portum salutarem dicto burgo sumptibus suis . . . . ad quem et in quo naues applicare valeant et in ipso applicantes ipsius maris fluxibus et refluxibus non obstantibus salue quiescere et stationem habere securam omni celeritate possibili edificabunt constructum edificatumque imperpetuum sustentabunt burgenses vero burgi supradicti in huiusmodi portus edificacione auxilium tale exhibebunt quod sumptibus suis lapides omnes sabolum et alia portus constructionem impediencia removebunt ac semel dictum portum purgabunt a sabolo et lapidibus cum pro dicto opere conueniens fuerit et necesse continuantes eciam dictam mundacionem a dicti portus fabrice inicio usque dictum opus fuerit

* The "Architecture" of M. Vitruvius Pollio ; translated by W. Newton architect. London, 1791. Vol. i. p. 121.

perimpletum archas omnes pro portu ordinatas ad consilium magistrorum implebunt et locabunt ac lapidibus onerabunt ad primam edificacionem portus supradicti instrumenta certa ad hec scilicet vangas, tribulos, et gavyllox ferreos eorum expensis inuenient alia vero instrumenta et onera Abbas et conuentus." Such is the account of the practice of Scotch engineers 450 years ago, from which it appears that the structures then erected consisted of timber chests filled with rubble stones.*

The earliest drawing of timber piers that I have met with is that of the ancient Port of Dunkirk in 1699, a copy of which, from Belidor's *Architecture Hydraulique*, is given in plate No. IX. This kind of structure is still not uncommon on the coasts of the English Channel.

Destruction of Timber by Marine Insects.—In sheltered bays, where a deep-water landing-place is all that is required, and where the bottom is sandy or soft, timber may be employed with great advantage. Even in somewhat exposed situations, it can also be used for tidal harbours; but the fatal evil in places where there is no admixture of fresh water, is its rapid destruction by marine insects. In the Atlantic Ocean, the *Teredo navalis* is very destructive to timber, and is not, as some writers suppose, of recent appearance on our shores. Upwards of 300 years ago, the ravages of these animals attracted much attention in Scotland, and gave rise to the most absurd theories as to their generation; and although the resemblance, one would think, is sufficiently obscure, they were then firmly believed to be young sea-fowl of the kind

* Professor Innes appends a point of interrogation to the word "*tribulos.*" On examining Agricola *De re metallica*, printed at Basil in 1561, and which contains descriptions of different implements, I found the *tribuli* described as a kind of *hammer* for breaking up ore. The tribuli were then obviously intended for breaking the stones with which the *archæ* or timber chests were to be filled.

called *Klaiks.** The barnacle (*Lepas anatifera*) was formerly supposed to be the young of the barnacle goose, as the modern scientific name indicates. The form of the barnacle shell has perhaps a faint resemblance to a pair of bird's wings, and there is a feathery plume, which the animal can thrust out, and which of course helps the resemblance. At many places in the German Ocean, the *Limnoria terebrans*, which was first discovered by the late Mr. Robert Stevenson at the Bell Rock Lighthouse in 1810, destroys most kinds of timber. Mr. Stevenson found † that Memel timber was destroyed by the Limnoria at the Bell Rock, at the rate of about *one inch inwards per annum.* At Kingstown, the *Chelura terebrans* has also proved very troublesome; and Mr. Rawlinson has found the common mussel nearly as destructive to timber as the Teredo. The Limnoria and the Teredo are found to eat most rapidly between the bottom and low-water mark, but above low-water the damage is not so great; and, what is singular, they do not appear to exist at all below the bottom, where the pile is covered with sand. This result does not, however, agree with Mr. Hartley's experience at Liverpool, where the parts which were alternately wet and dry were noticed to decay faster than those which were constantly immersed.

Experiments at Bell Rock.—Mr. R. Stevenson devoted much attention to the ravages of the *Limnoria terebrans* at the Bell Rock, where he established a regular series of observations, beginning in 1814, which were made by fixing pieces of different kinds of timber to the rock, and getting regular reports on their decay. He found that *greenheart, beef-wood,*

* Hector Boece's "Kroniklis of Scotland," published at Edinburgh about 1536.

† Account of the Bell Rock Lighthouse, by R. Stevenson, F.R.S.E Edinburgh, 1824.

African oak, and *bullet-tree*, were scarcely attacked by worms while *teak* stood remarkably well, and *locust* tolerably well although suffering at last.* He subsequently made other experiments, the results of which are also appended in the following Table :—

* Dr. Bancroft describes *Greenheart* or the Sipiera tree to be "in size like the locust tree, say 60 or 70 feet high : there are two species, the black and the yellow, differing only in the colour of their bark and wood. The greenheart of Jamaica and Guiana is the *Laurus chloroxylon* of botanists ; it is also called cogwood in the former, and Sipieri in the latter locality." . . . .

"Botany Bay Oak, sometimes called *Beef-wood*, is from New South Wales ; it is shipped in round logs from 9 to 14 inches diameter. . . . *Casuarina stricta* is called she oak, and also beef-wood." . . . .

"The *African oak*, or *teak*, as it is called, is not a species of *Quercus*. . . . *Teakwood* is the produce of the *Tectona grandia*, a native of the mountainous parts of the Malabar coast, and of the Rajahmundry Circass, as well as of Java, Ceylon, and the Moulmein and Terrasserim coasts. . . . . In 25 years the teak attains the size of 2 feet diameter, but it requires 100 years to arrive at maturity. . . . . African teak does not belong to the same genus as the Indian teak ; by some it is thought to be a *Euphorbiaceous* plant, and by Mr. Don to be a *Vitex*." . . . .

"*Bullet-wood* from the Virgin Isles, West Indies, is the produce of a large tree with a white sap ; the wood is greenish hazel, close and hard. It is used in the country for building purposes, and resembles the greenheart. . . Another species so called is supposed to come from Berbice."

"*Locust-tree.*—The locust-tree of North America is *Robinia pseudacacia*. The wood is greenish-yellow, with a slight tinge of red in the pores ; it is used like oak. Locust is much esteemed for treenails for ships. . . . The locust-tree of the West Indies and Guiana is *Hymenea Courbasil* (Somiri), a tree from 60 to 80 feet in height, and 5 or 6 feet in diameter. The colour of the wood of the West India locust-tree is light reddish-brown, with darker veins, and the mean size 36 inches."—*Descriptive Catalogue of the Woods commonly used in this country, with Botanical Notes*, by Dr. Royle : London, 1843.

TABLE showing the different kinds of Timber which were exposed to the attacks of the *Limnoria terebrans* at the Bell Rock in 1814, 1821, 1837, 1843, with their durabilities.

Kind of Timber.	Decay first observed.	Unsound and quite decayed.	Quite sound for	Remarks.
	yrs. mo.	yrs. mo.	yrs. mo.	
[1]Greenheart	..	..	19 0	[1] Affected in one corner.
Teakwood	..	..	13 0	
Beef-wood	..	..	13 0	
Treenail of Bullet-wood	..	..	5 0	
[2]Beech, Payne's patent pro.	10 7	..	..	[2] A little holed at one end underneath.
[3]Teakwood	5 6	..	..	[3] Nearly sound 7½ years after being laid down.
[4]African Oak	5 6	..	..	[4] Nearly sound 7½ years after being laid down.
Do. do.	4 11	10 0	..	
English Oak, kyanised	4 7	10 0	..	
Teakwood	4 7	12 0	..	
[5]American Oak, kyanised	4 3	..	..	[5] Decaying but slowly 5 years and 7 months after being laid down.
British Ash	3 0	5 0	..	
Scotch Elm	3 0	5 0	..	
Ash	2 11	4 3	..	
English Elm	2 11	4 7	..	
[6]Plane Tree	2 11	..	..	[6] Decaying but slowly 5 years and 7 months after being laid down.
American Oak	2 11	4 7	..	
[7]Baltic Red Pine	2 9	4 3	..	[7] A good deal decayed when first observed.
English Oak	2 4	4 7	..	
[8]Scotch Oak	2 4	..	..	[8] Much decayed when first observed.
Baltic Oak	2 4	4 3	..	
Norway Fir	2 4	3 1	..	
Baltic Red Pine, kyanised	2 4	4 7	..	
Pitch Pine	2 4	4 3	..	
American Yellow Pine	2 4	3 7	..	
American Red Pine	2 4	3 1	..	
Do. do., kyanised	2 4	4 7	..	
Larch	2 4	4 3	..	
[9]Honduras Mahogany	2 1	..	..	[9] Nearly sound 3½ years after being laid down. Washed away 6 months later.
Beech	1 9	3 1	..	
American Elm	1 9	3 1	..	
Treenail of locust	..	5 0	3 0	
British Oak	1 6	5 0	..	
American Oak	1 6	5 0	..	
Plane Tree	1 6	5 0	..	
Honduras Teak treenails	1 6	5 0	..	
Beech	1 6	5 0	..	
Scotch Fir, teak treenails	1 6	3 0	..	
Do. from Lanarkshire	1 6	3 0	..	
Do. do.	1 6	3 0	..	
Do. locust treenails	1 6	3 0	..	
Memel Fir	1 6	5 0	..	
[10]Pitch Pine	1 6	2 6	..	[10] Going fast when first observed.
English Oak	1 1	3 1	..	
Italian Oak	1 1	3 6	..	
Dantzic Oak	1 1	2 6	..	
English Elm	1 1	1 6	..	
Canada Rock Elm	1 1	1 6	..	
Cedar of Lebanon	1 1	2 6	..	
Riga Fir	1 1	1 6	..	
Dantzic Fir	1 1	1 6	..	
Virginia Pine	1 1	1 6	..	
[11]Yellow Pine	1 1	1 6	..	[11] A good deal gone 18 months after being laid down. Swept away by the sea 7 months afterwards.
Red Pine	1 1	1 6	..	
[12]Cawdie Pine	1 1	1 6	..	[12] A good deal decayed when first observed.
[13]Polish Larch	1 1	1 6	..	[13] Going fast when first observed.
Birch, Payne's patent pro.	0 10	1 10	..	
American locust treenails	0 8	3 0	..	

Greenheart timber, though not absolutely impenetrable, as appears from Mr. Stevenson's experiments, is the great specific in seas where the worms are destructive. Greenheart appears to have been first used as a material by Mr. J. Hartley, who, in 1840, published, in the Minutes of Institution of Civil Engineers, an account of its virtues, as ascertained at the Liverpool Docks. Its cost is, however, considerably greater than Memel, or than most of the other timbers in common use. Mr. D. Stevenson gives the following account of recent experience:—"It was, I believe, for the first time employed for staging at Wick Bay, where logs of pine could not withstand the waves; and it was on removing the temporary greenheart staging, that had been in use from two to four years at Wick, that I first became fully aware that the Limnoria would perforate that timber. Some of these logs were found to have been attacked by the Limnoria throughout the whole surface, extending from about low-water mark to the bottom. This discovery caused no little surprise and regret, as engineers had always looked on greenheart as proof against destruction by marine insects; but being the first, and it was hoped perhaps an isolated instance, I did not consider it necessary at once to record the fact.

"I have since, however, received a specimen of timber taken from one of the piles in the steamboat pier at Salen, in the Sound of Mull, which was erected four years ago, the main piles being made of sound greenheart, and I find that in this locality also the Limnoria has commenced to perforate the timber.

"In both of these instances sufficient time has not elapsed to allow the wasting to make great progress, but in both cases the perforators have penetrated into what is unquestionably sound fresh timber; and therefore this result conflicts with certain other experiments, such as those made at the Bell Rock, where the greenheart remained nearly sound after nineteen years' exposure.

"The joint paper of Dr. Maclagan and Dr. Gamgee on greenheart in the Society's 'Transactions' states that by subjecting greenheart *wood* to a process identical with that used for the extraction of sulphate of bebeerine from the *bark*, a product is obtained possessed of an intensely bitter taste, and not differing perceptibly from the sulphate of bebeerine. This may account for wounds produced by a splinter of greenheart not readily healing.

"I am also disposed to think that it is to the existence of this alkaloid in the timber, and not to its hardness, that its undoubted power of withstanding, in certain cases and for a certain time, the action of the Limnoria is due; and it would be interesting to discover whether the wasted portions of greenheart at Wick and Salen produced bebeerine in a smaller degree as compared with sound timber. It is possible, as suggested by Sir Robert Christison, that long-protracted immersion in sea-water may so counteract the preservative principle due to the bebeerine in the timber as to render it open to attack. It is also possible that the greenheart now imported in such large quantities has degenerated, like the 'Crown Memel,' which, it is well known, cannot be procured of the same high quality as formerly. Change of soil, moreover, affects the growth of trees, and is perhaps sufficient to account for the great variations in the quality of foreign-grown timber.

"In any view of the case, however, it seems necessary, in connection with my former notice, to make known the fact that greenheart, *as now imported*, and generally used in marine works, is not, as was hitherto supposed to be the case, wholly proof against the ravages of the *Limnoria terebrans*, suggesting, perhaps, increased care in its selection, although I believe it must still be regarded as the most durable timber that can be employed in such works. It is almost unnecessary to add that these observations refer to localities where the timber is exposed to what may be termed *sea-water*, and not to situations where, from admixture of fresh water or other causes, the ravages of the Limnoria are greatly mitigated, or altogether unknown."*

Protection of Timber.—Memel logs for the interior piles of piers, where they will not be liable to suffer by abrasion from ships, might perhaps be clad with greenheart planking at those parts which are exposed to the worm. *Copper sheathing and scupper nailing* are often successfully employed as protections for piles in exposed situations. The scupper or broad-headed nails are driven so closely as almost to touch each other; and the oxide of iron enters into the outer skin of the wood, which becomes hard enough to defy the worm. *Green twigs of pine or other wood*, when placed among piling, have been found, in Sweden, to prevent the attacks of the worm. *Breaming* or scorching timber, and saturating it, while hot, with a mixture of whale oil and thin coal tar, also forms a temporary protection. The *creosoting process*, patented by Mr. Bethel, and which has been so largely and so successfully introduced for preventing the decay of railway sleepers, bridges, etc., has recently been much employed in the construction of timber works subject to the attacks of marine insects. Mr. Bethel recommends that the timber for such purposes should receive 10 lbs. of creosoting fluid to each cubic foot, which is tested by weighing each log before and after it leaves the creosoting tank. It was confidently be-

* Proceedings Roy. Soc. Edin. vol. viii. p. 781.

lieved that Mr. Bethel's very important invention, which had proved so efficacious in preventing the ordinary decay of wood on land, would be found equally useful for timber immersed in sea-water. It was first found, as might indeed have been readily anticipated, that it would not answer the end expected, if the timber were cross cut, or scarfed, *after* the fluid had been injected. But it was afterwards discovered that the woody fibre was eaten, even although the outer skin had suffered no injury of any kind after being in the tank. Mr. D. Stevenson has lately directed attention to this fact, and has proved that at Scrabster, Invergordon, and other places where the timber was thoroughly creosoted, it has been very much destroyed by the worm, which undoubtedly eats the timber freely, even though it be still black with the creosote, and continue to emit its pungent odour. Mr. A. M. Rendel, in his evidence on Leith Harbour, asserts, from the experience he has had at that port, that the life of timber fully creosoted is limited, at Leith Harbour, to about 20 years.*

Pile-work.—Dr. Rankine's formulæ † for the strength of pile-work are, when—

P is greatest load which a pile is to bear without sinking farther, in tons;
W, the weight of ram used for driving it, in tons;
h, the height from which the ram falls, in feet;
l, the length of the pile, in feet;
x, the depth it is driven by the last blow, in fractions of a foot;
S, its sectional area, in square inches;
E, its modulus of elasticity;
(Approximate values of E in tons on the square inch—elm, 400 to 600; beech, about 600; greenheart, 500 to 600)—

* Evidence, Select Committee on Leith Harbour. 1860.

† Useful Rules and Tables. By W. J. M. Rankine. Lond. 1866, p. 183.

$$x = \frac{Wh}{P} - \frac{Pl}{4ES}.$$

The pile must be driven until the additional depth gained by each blow of the energy Wh becomes not greater than x.

The energy required for the final blow is—

$$Wh = \frac{P^2 l}{4ES} + Px.$$

And, finally—

$$P = \sqrt{\left(\frac{4ESWh}{l} + \frac{4E^2S^2x^2}{l^2}\right)} - \frac{2ESx}{l}.$$

Professor Stevelly gives a simpler formula, which assigns a considerably smaller value to the safe load.

When W is weight of ram, in tons;
W′, weight of pile, in tons and decimals;
h, height of fall, in feet and decimals;
d, depth yielded, in feet and decimals;
L, safe limit of load, in tons;

$$L = W\left(\frac{W}{W + W'}\right)\left(\frac{h}{d}\right).$$

Screw-Piles.—The load supported by a screw-pile in practice ranges, according to Rankine, from 3 times to 7 times the weight of the earth which lies directly above the screw-blade.

Advantages of Timber over Stone as a Material.—It is much to be regretted that greenheart, which so long resists the worm, is so expensive in this country, and that some simple and economic specific against the worm has not been discovered for protecting Memel and the cheaper kinds of pine. The grand desideratum in harbour works, which is the *want of continuity in the structure,* would be supplied by timber work. It follows from the known laws of fluids that each individual stone in a pier which is equally exposed throughout its whole length, is subjected to a force which it can only resist by its own inertia, and the friction due to its contact with the adjoining stones. The stability of a whole work, if not cemented by hydraulic mortar, may therefore be perilled by the use of small stones in one part of the fabric, while it may be in no way increased by the introduction of heavier stones into other parts. By the use of long logs of timber, carefully bolted together, a new element of strength is obviously obtained.

Destruction of Stone.—Even solid rock is destroyed by the persevering efforts of the Pholades and Saxicavæ. The Pholas perforates wood, limestone, hard and soft argillaceous shales, clay, and sandstone. Though the Saxicavæ, which attack the limestone blocks of the Plymouth breakwater, do not penetrate more than half a foot from the surface, yet their holes are so close to each other as to make it easy to break off the outer portions of the stone, when a new surface is laid open to their attack.*

Destruction of soft Rock in situ *by the Pholas.*—I observed lately at Kirkcaldy a curious example of very serious mischief which had been caused by the gradual excavations of the Pholas. The quay-walls and gate-chamber of the scouring-basin, which is also used as a wet dock, were built on beds of shale or tile, which I am told formed originally a most secure and incompressible foundation. But several years after the work was finished the masonry gradually settled, and is now so much sunk that both the quays and gate-chamber have become ruinous. Persons on the spot believed that the settlement was occasioned either by want of strength in the masonry, or by the sinking of old coal workings, which were supposed to exist below the harbour. But on examining the bottom, I found it completely honeycombed by the Pholas,

* History of British Mollusca. By Edward Forbes and S. Hanley. Lond. 1853. Vol. i. p. 104.

which, getting access through the water in the dock, had perforated the shale-beds on which the walls rested, and which, before the dock was excavated, had never been exposed to their attacks. So firm and compact had the shale been at first that the masonry, instead of being carried down to the level of the bottom of the dock, was founded on the top of the shale.

As the number of perforations in the shale increased, its power of resisting compression must obviously have gradually decreased in a corresponding degree, till at length the weight of the quay-wall would begin to crush the shale. But it is evident that the settlement thus occasioned could not have been equal over the whole area of the foundation, because the outer portions of the shale, being next the water, would necessarily be more honeycombed than the interior, and hence the outer facing of the walls would sink more than the *backing,* which was precisely what was found to take place.

Iron.

Mr. R. Stevenson's Experiments on the Durability of Iron.—In addition to the experiments on timber, twenty-five different kinds and combinations of iron were tried at the Bell Rock, including specimens of galvanised irons. All the ungalvanised specimens were found to oxidise with much the same readiness. The galvanised specimens resisted oxidation for three or four years, after which the chemical action went on as quickly as in the others. Although the association of zinc with iron protects, so long as it lasts, the metal with which it is in contact, it must be remembered that this immunity is obtained at the expense of the zinc, the tendency of which to oxidisation is proportionally exalted so soon as any part of the iron is exposed.

Mr. George Rennie's Experiments on the Durability of Wrought-iron, Cast-iron, and Bronze.—Mr. George Rennie made experiments in 1836 on one-inch cubes of wrought-iron, of cast-iron, and of bronze, with reference to the question of their eligibility for lighthouse purposes. In narrating his experiments and their results Mr. Rennie says—

"The cubes, being previously weighed, were then plunged into a saline dilution considerably stronger than sea-water, as follows:—

Muriate of Soda	122	grains.
Muriate of Magnesia	25	"
Muriate of Lime	6	"
Sulphate of Soda	30	"
	183	grains dissolved in $10\frac{1}{2}$ oz. of Thames water.

"The cubes were then taken out of the water, after being immersed seventy hours in separate vessels. The cast-iron was found to have lost $\frac{1}{3307}$th part of its weight, while the wrought-iron had only lost $\frac{1}{6810}$th of its weight, being in the proportion of *two lost by the cast-iron to one only lost by the wrought-iron;* while the brass cube only lost $\frac{1}{10000}$th part of its weight, which is decisively *in favour of bronze, in the ratio of three to one.*

"The cast and wrought iron cubes, being accurately weighed, were again plunged into a strong dilution of 1 measure of muriatic acid to 25 measures of Thames water, when, after remaining twenty-one hours, the cast-iron cube was found to have lost $\frac{1}{33}$d of its weight, and the wrought-iron only $\frac{1}{258}$th of its weight, being in the proportion of 8 *to* 1 *in favour of wrought-iron.*"

Ancient Bronze Relics.—The wonderful durability of bronze is well shown by the axe-heads and other articles belonging to prehistoric periods, which are from time to time discovered in making excavations in gravel-drift. Five bronze axes were recently turned up in the works of excavation for the Edinburgh and Leith sewerage. Nothing could exceed the sharpness of the edges and the projections of the ornamentation, and they were all remarkably free from oxide.

Mr. Mallett's Experiments.—The important experiments of Mr. Mallet on specimens sunk in the sea showed that the amount of corrosion *decreased with the thickness of the casting, and that from $\frac{1}{10}$th to $\frac{4}{10}$th inch in depth, in castings* 1 *inch thick, and about $\frac{6}{10}$th inch of wrought-iron, will be destroyed in a century in clear salt water.**

Examples of very rapid Decay.—There are in my possession specimens which prove that with some kinds of iron the rate of oxidation in thick castings had been more rapid than in the samples employed in Mr. Mallett's experiments. In 1833 a cannon-ball, $4\frac{1}{2}$ inches diameter, was picked up on the eastern shore of Inchkeith island, in the Firth of Forth, at a place that was left dry two hours and a half before low water of spring tides. It was, therefore, not constantly immersed, but was only alternately wet and dry, a condition which is generally believed to retard materially the progress of decay. The shore at the place is gravelly, with rocks intervening, so that there was no peculiarity of soil that could have hastened chemical action. The external appearance of the ball is precisely that of any ordinary casting which had for a long time been exposed to atmospheric influence, indicating the presence of red carbonate in some parts, while in others there appears a smooth skin, possessing a certain degree of metallic lustre. The raised ring which had been formed by the edges of the mould is still quite apparent, and the radiated structure of the interior is also distinctly visible. But so thoroughly has

* Brit. Association Report, 1839 and 1850.

the iron been changed, that the ball weighs only *one-fourth* of what it would had the metal remained sound. So perfect is the transmutation, that there cannot be detected, even in the centre, the slightest trace of unaltered metal. The whole forms an earthy substance, consisting, I believe, principally of carbonate of iron.

The preference which is now so frequently given to iron as a material, even in cases where a nearly indestructible substance ought, if possible, to be employed, makes it of some consequence to ascertain the probable date of the immersion of this cannon-ball, which there is every reason to believe from the following facts was not earlier than 1564.

The first use of cannon in warfare is commonly believed to have been at the battle of Cressy in 1346; but as it is known that balls of stone were first employed, it seems very improbable that well-formed balls of cast-iron were made until long after that period. The account of the earliest military operations at Inchkeith that I have seen is given by Sir Robert Sibbald in his History of Fife and Kinross, who states that, in the reign of Elizabeth, its capture became a matter of keen contest; so much so, that the English had then a fleet of twenty-nine vessels anchored off the island. The remains of Queen Mary's fort, which was erected for her by the French, and which bears her initials, with the date 1564, may still be seen on its summit. There is therefore good reason for supposing that it was during those contests, *at soonest*, that the ball found its way into the sea. If this be so, it proves that 2¼ inches of cast-iron (the radius of the ball) became thoroughly oxidised in the space of not more than three centuries, which assigns for the kind of iron of which it was cast *upwards of ¾ inch to the century for balls 4½ inches diameter*, and placed so as to be alternately wet and dry.

At the Bell Rock Lighthouse, which was completed in 1810, Mr. R. Stevenson directed cast-iron tramways, consisting of rails with open gratings between, and supported on standards, to be fixed to the rock. Many of these gratings, which are *not* constantly immersed, are now decayed in different places, cavities having been formed on their upper surfaces fully half an inch deep, thus giving *one inch to the century* for castings an inch square. It is remarkable, however, that in some of the specimens of this grating which I have examined, the decay is principally confined to those parts where there have been air-holes in the metal. The rapid decay in these holes is probably occasioned by the water being retained in them after the tide has receded; and thus the increased action due to constant immersion is produced at that part of the casting. Where iron is to be exposed to only periodic immersion, it therefore comes to be of especial importance that the castings should be not only free from air-holes, but should be of a perfectly regular and slightly rounded form, so as not to present hollows for the water to lodge in. One of the bars, however, which was quite free from air-holes, and presented no external appearance of decay, had its specific gravity reduced to 5.63, and its transverse strength reduced from 7409 to 4797 lbs. Another apparently sound specimen was reduced in strength from 4068 lbs. to 2352 lbs., *having lost nearly half its strength in about fifty years;* and I strongly suspect that all the gratings, however sound they may look, have suffered a great reduction of their strength.

Even although the instances of such rapid oxidation as have been adduced be of rare occurrence, yet the bare possibility of such a speedy decay should discourage the indiscriminate employment of iron in marine works. Where there is room for choice, neither cast nor malleable iron should be used as *principal* constituents of any structures which require to be so deeply submerged as to become difficult of inspection or repair. If even thick castings, and those not constantly immersed, decay so rapidly, what can be expected of the durability of malleable iron bolts and tie-bars which do not exceed an inch or two in diameter? And what reliance can be placed on the stability of deeply submerged structures retaining large quantities of rubble, the unity of which depends wholly on such perishable bonds?

Dressing and Method of assembling Masonry in Sea-works.

For localities where the exposure is not very great, and where cement cannot be procured, masonry consisting of separate blocks may be adopted.

The requirements of marine masonry are, in many respects, nearly the opposite of those for land architecture. What is wanted in sea-work of the ordinary kind, which neither consists of framed carpentry, nor has been rendered monolithic by the use of cement, is that each stone shall gravitate freely, and transmit its pressure unimpaired to those below it. If, therefore, a pier could be so constructed, that on the abstraction of a stone at the bottom, the whole vertical section of masonry resting upon it should at once sink, so as to fill up the void, the perfection of marine masonry would then be attained, because the lower courses would bear the unreduced weight of the upper, and would therefore be the less easily abstracted. The difficulty of pulling out any stone in such a work would then be proportional to its distance from the top of the wall. Whereas in land architecture, vertical bond is systematically preserved, and the stones are sometimes so lintelled over, or, to speak technically, so completely "*saved*" from the superincumbent pressure, that it is often easy to extract some of the lowest stones in the wall, without endangering the stability of the upper courses. In land architecture, the whole structure is also greatly strengthened by the occasional insertion of long headers and stretchers, but in the sea, where each stone is assailed *per se*, the stability of any horizontal course, if equally exposed throughout, is measured by the stability of the *smallest* stone in that course. And, therefore, the more uniform the size of the materials in each horizontal section of the work the better, provided that the "secret-bond," or proper connection with the backing, is duly preserved. We must beware then of importing into marine engineering, as is too generally done, the laws and maxims of house-architecture, with its careful vertical bond, and its small but finely dressed face-stones. It matters not, indeed, how rough the masonry of the *face-work* of a pier be, provided there are no protuberances large enough to offer material resistance to the jet of water in front of the wall; and we have already pointed out the valuable effects of keeping the beds rough. All the blocks should, however, *bed and joint* fairly on each other, and no face pinnings, or small *closers*, should on any account be allowed in the outer face-work. It is also of vital consequence, that the *backing* should not be slurred over, by being loosely assembled; but should, on the contrary, be carefully set, and regularly bonded with the face-work as the building proceeds. The *outside* of the *parapet*, though of smaller dimensions, should be similar in quality to the sea-wall, while its inside, from not being exposed to the wash of the sea, may be built of good heavy rubble. The whole parapet may with advantage be set in mortar.

Roadway Pitching.—From the risk of damage already

referred to, it comes to be a difficult question to decide whether the *roadway pitching* should be built with very open joints, or be made altogether impervious to water. Mr. T. S. Hunter, in his report on the Wansbeck River, mentions the following instance of damage to the pitching at Granton. "A portion of the pitching, which had just been grouted previous to the storm, was completely doubled up like a sheet of paper, but after the grouting was removed, and the same stones set dry, they were never again disturbed." The safe course in most cases, where the sea-wall is built of dry masonry, is probably not to attempt the formation of an altogether impervious covering; but where the roadway is made impervious, the compressed air may be discharged through the mooring-pauls.

The *quay-wall* requires no particular notice. Minard recommends its thickness to be $\frac{4}{10}$ths of the height. The upper portions are sometimes set in mortar, but the rest is set dry. The *rubble hearting* should be free of earthy or clayey matter, or rock of a quality likely to crumble on exposure. Very large boulders ought not to be admitted, unless after being broken up, and when any of the stones are long and flat, they should be laid lengthways of the pier, and in no case, unless when the structure is of great width, should the stones be tipped in without being afterwards carefully assembled with the hand.* The *rubble for breakwaters* is generally of a much larger size than for ordinary commercial piers. The *ratio of voids* in a cubic yard of rubble after being deposited depends of course on the kind of materials, and

* In exposed situations or in narrow piers, the hearting, which may be regarded as the *back-bone* of the work, should not only be carefully assembled, but the stones should be set hard on each other, so as to give a continuous bearing throughout the whole width of the work.

has been found to vary generally from about 4 to about 8 or even 9 cubic feet in each yard, equal to from $\frac{1}{9}$th to $\frac{1}{3}$d of increased cubic space in the breakwater.

Settling of Rubble.—The materials of breakwaters are liable to sink, to be crushed, and to be driven together by the waves. At Cherbourg, settlements of about 4 inches were observed to take place after a tempest. At Algiers, where the bottom is soft, the rubble sank 6.5 feet into the sand, and at Boyard 5232 cubic yards were required below the level of the bottom. According to the late Admiral Washington, the settlement at Cherbourg* averaged 18 inches in 22 feet, or *one-fourteenth* of the height.

Edge Work.—The method of assembling stones on edge, instead of on their beds, which was used in some old Scottish harbours and sea-walls, as at St. Andrews, Prestonpans, etc., deserves to be more generally known and adopted, from its greatly superior strength. Mr. Bremner of Wick propounded the opinion that—"If the walls are constructed on a (horizontal) angle of 25° to the sea, and the materials built on edge with 3 inches of slope to the foot perpendicular, they cannot retain any air, and the sea running along a small portion of the building at one time, actually assists in forcing together the edge building." Although it amounts virtually to a condemnation of nearly all modern sea-work, yet I do not hesitate to assert that it is a great engineering error to assemble stones in exposed works in any other way than on their edges, and I extend this remark even to materials of ordinary thickness, although the advantage is most conspicuous where the materials are thin. Care must be taken, however, not to adopt this plan where there is any risk of heavy seas coming in a wrong direction, so as to strike the masonry on the overhanging side.

* Minard, pp. 56, 59.

Rhomboidal Form of Stones.—The strength of the masonry of sea-walls might be materially increased if the stones were dressed to a rhomboidal form. This could be done without much trouble or expense, if the quarriers were furnished with bevelled templates, instead of square, for quarrying the blocks.

Size of Materials.—In absence of any rule for the weight of materials, the following approximation is submitted rather for trial than as a guide. When W = average weight of blocks in tons, and d = length of fetch in miles, $W = .3\sqrt{d}$.

Treenailing.—A most efficient method of temporarily increasing the stability of marine masonry, which may be adopted in places where the materials are of small size and of a soft texture, is to connect all the stones together by a system of dowelling or treenailing. Each block is thus secured to its neighbour by iron bolts or wooden pins let into the lying beds, or into both beds and joints. At the Eddystone and the Bell Rock lighthouses, the stones were not only secured by oaken treenails, but were also cut so as to dovetail into each other, and thus to render the mass practically monolithic. They were also further secured by vertical wedges. But these methods are attended by an expenditure which is warrantable only in peculiar works like those, where the loss of a single block is certain to occasion great inconvenience and delay. Mr. Leslie used timber treenails largely at Arbroath and Kirkcaldy, where the stones for the works were freestone. The holes were bored by means of machines made for the purpose, at far less expense than if the ordinary tedious process of hand "jumping" had been adopted.

Beton.—Beton is now very commonly used in this country, as well as on the Continent. This most valuable material may be said to have, to a large extent, revolutionised harbour-building : for it admits of being employed in many different ways, and can, if due care be taken, be used with perfect confidence. Sir John Hawkshaw's specification I have used extensively; it is as follows:—

"The Portland cement is to be of the very best manufacture, from the Thames, or other Portland cement of equal quality. It is to be ground extremely fine, and is to weigh not less than 115 lbs. per bushel, and each cargo will be tested in the following manner:—The cement is to be made into small blocks of a convenient size, and after forty hours the blocks are to stand a tensile strain of not less than 112 lbs. per square inch, and after sixty hours they are to stand a tensile strain of not less than 150 lbs. per square inch. Moreover, the blocks are to be immersed in water, and after seven days they are to stand a tensile strain of not less than 200 lbs. per square inch. Slabs or cakes are also to be made, and placed in water, and after immersion for twenty-four hours they are not to show any signs of cracking, or any softness on the surface. No cement is to be used until it has been in the sheds for at least one month, unless the engineers shall otherwise permit, and they will, if they think fit, order this time to be lengthened."

"*Concrete.*—The concrete will be composed of one measure of Portland cement, four measures of sand, and five measures of shingle. The concrete is to be made in the following manner:—The materials, having been measured, are to be well mixed with a due proportion of water, and the whole well worked together and put into the work immediately it is mixed."

Blocks of Cement Rubble used at Wick.—My friend Mr. Alan Brebner, who took charge of the arrangements for depositing blocks of cement rubble at Wick, has kindly drawn up the following particulars:—The large blocks constructed for the protection of the end of the breakwater having to be floated into their places by means of machinery placed upon lighters, were made in positions within the high-water mark, such as would permit of the lighters floating alongside of them at any tide. The range of spring tides being only about 10 feet, sites were selected on the beach about 2 feet above the level of low-water spring tides. The first operation was to prepare level beds for the timber platforms on which the blocks were built. The platforms measured 22 feet long and 16 feet broad each, and consisted of 4 cross-sleepers 12″ × 6″, bolted down to 12 stones, about 15 cwt. each, previously sunk into the beach and levelled. The sleepers were covered with 3-inch planking, jointed with Roman cement, driven close together, and then spiked down. Twelve platforms were constructed, at a cost of about £20 each. The sides of the boxes were formed of 3-inch planks, secured at the corners by palm bolts and nuts, so as to admit of their easy removal after a block had been completed. The joints of the planks were slightly bevelled to the outside, leaving a wedge-shaped space, which was filled with Roman cement to exclude the water. The blocks were all 5 feet 6 inches high, and this height was made up in 6 widths of planking. Three widths of planking were put down, and filled, generally in one tide, the other three in the tide following, and the block completely finished and smoothed off in the third tide. The blocks, which weighed from 80 to 100 tons, consisted of one part of Portland cement to seven of small stones, sand, and gravel; they were not what is commonly termed built blocks, the whole of the stones being assembled by labourers only. The stone used was of a high specific gravity, and in pieces of from 10 to 80 lbs. weight.

In commencing the blocks a covering of stones was first laid over the bottom, care being taken that spaces nowhere less than one inch was left betwixt them. The cement, sand, and small gravel were made into a thinish mortar, and conveyed into the box in wheelbarrows. The spaces betwixt the stones were completely filled, and a covering of 2 or 3 inches laid over them to form a bed for those to follow. A sufficient force of men was always employed to build a block half up in one tide, and before the water covered it a sheet of canvas was laid over it, and loaded with stones, to prevent the waves from acting on the cement until it had time to harden. Next tide the covering was removed and the block completed.

Near the four corners of the blocks, boxes were inserted to form holes for the lifting bars. Those boxes rose one foot up into the blocks, and from that level tapered battens were carried up above the top, and withdrawn when the cement had set. Bars of iron 3 in. × 1 in. were built into the holes to prevent the lifting bars from cutting into the cement work. In two days after the blocks were finished the wood was stripped off them, and in eight days they were fit to be, and sometimes were, lifted from the platforms and built into the pier. A much smaller proportion of cement is generally used in making concrete blocks, but the unusually heavy seas in this situation required the blocks to be very strong.

Lifting of the Blocks.—The machinery used for lifting and setting the blocks is shown in the diagrams Figs. 27 and 28, and consisted of two lighters, each capable of carrying 50 tons; on these were erected two strongly-trussed timber frames, tied and braced together, so as to preserve the lighters in a position parallel to each other. Two logs of greenheart timber rested on the trusses, and carried the brackets and pulleys for guiding the chains to the winches, and also the two upper pulleys of the lifting tackle. The blocks of concrete being of a considerable size, it was considered advisable to lift them by four points, to avoid the risk of breaking in the lifting, and also to distribute the load as much as possible over the floating structure. It was evident that a strain considerably more

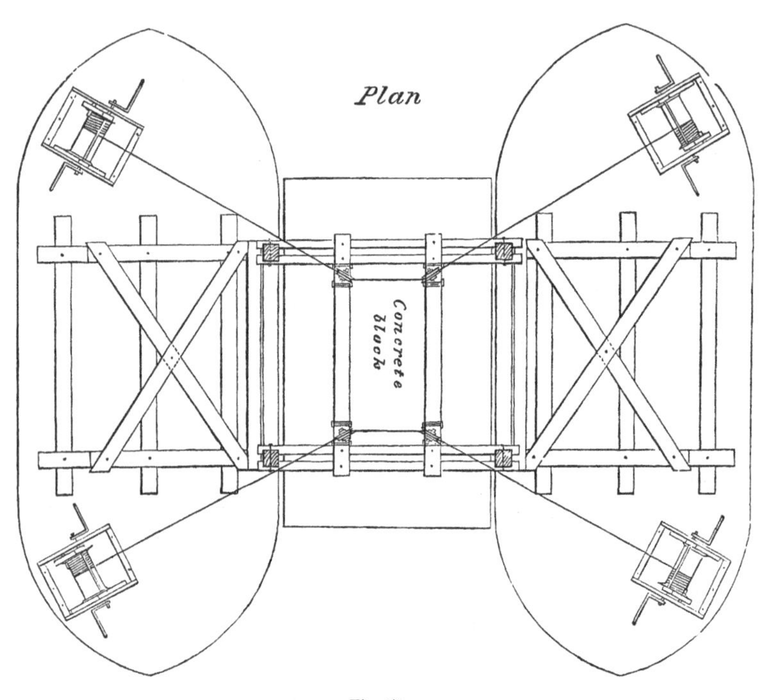

Fig. 27.

than what was due to the dead weight would be thrown upon the machinery by the motion of the waves, which sometimes broke over the seaward lighter, and fell on the top of the blocks betwixt them, adding greatly to the weight.

In order to secure an equal strain on each of the four lifting bars, which could not have been accomplished if four separate tackles had been employed, the chains were so arranged that what is usually the standing part of the one tackle was made the hauling part of its neighbour, two pieces of chain thus forming the four sets of tackle. The arrangement will be readily understood by referring to the diagrams. This arrangement worked admirably, the friction in the pulleys being more than sufficient to provide against any want of balance arising from the inaccurate placing of the lifting bars : it permitted the blocks being lowered into their places by means of breaks, which saved a great deal of time, and very much facilitated the setting. It also obviated all danger of

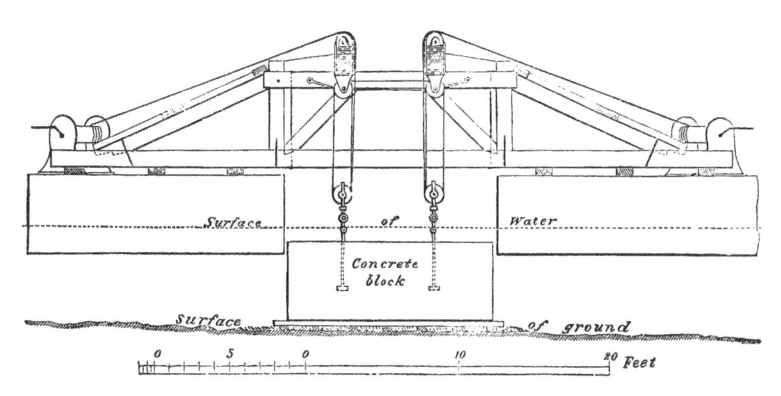

Fig. 28.

the tackles breaking from the work intended for two being at any time suddenly thrown upon one. The tackles were worked by four powerful winches arranged as shown on the diagrams, and worked by four men. When a block was to be lifted, the lighters were warped into the position shown in Fig 27, the lifting bars inserted, and the winches set to work. In four or five minutes the block was clear of the ground, and was then warped out, and either deposited in deep water to make room for others being built, or taken out and set in the work. To enable the blocks to be lowered down in a level position, the following expedient occurred to me, and was adopted with satisfactory results :—The winches were numbered, and each had its number painted up in a position to be seen by the men working it, and also by the person directing the operations. After the tackles were mounted, and the chains wound upon the barrels of the crab winches, so that the whole was in the position shown by Fig. 28, white marks were drawn across the framing of the winches and large spur wheels on the end of the winding barrels. By this means the person in charge, who stood on a platform on the top of the framing, was enabled to observe, during the process of lowering, the approach of the marks on the wheels to those on the framing, and thereby to adjust the level of the blocks at each revolution, and so ensure their reaching the bottom in that position. The lighters were steadied over the work by ropes attached to four mooring buoys laid down in the required positions. When nearly over their places, the blocks were lowered to within a few feet of their beds, and when finally adjusted, were lowered quickly to their places on a signal from a diver. The time taken to set a block depended greatly on the state of the sea. When quite smooth the operation was completed very quickly, at other times it would occupy about an hour.

The whole process, both of constructing the blocks under the tide level, and floating them into their places, was perfectly successful, and was carried out without the slightest accident or breakage : the first block was lifted with perfect ease, and no change of any kind was required on the machinery as at first constructed.

Voids between Blocks of Beton.—"Sometimes," says Minard, "the workmen succeed in placing the blocks close together;

but generally there are spaces between them of from $\frac{6}{10}$ths of a foot to 1.6 foot, and sometimes 2.6 feet. At Algiers, with blocks of 353 cubic feet (11.15 feet long × 6.56 broad) *thrown down on each other*, the voids are *one-third*." *

Passing Concrete through Water.—Sir John Hawkshaw† has passed concrete through 50 feet of water with perfect success. As far as his experience went, the concrete set quite as well under these circumstances as when it was deposited in the open air. He has done this both in salt and fresh water. In passing concrete through water, he used a box containing about two cubic yards. When it reached the bottom, a bolt was withdrawn, and the concrete dropped out. At the harbour of Greenock, Messrs. Bell and Miller put in the low-water foundation of the Albert Quay in concrete of Arden lime. Mr. Pascal, the engineer of the Marseilles works, formed in concrete the foundations of a beacon below low water at a submerged rock five miles from the shore. Mr. W. Parkes put in the foundations of the iron lighthouse in the Red Sea by means of a caisson, into which fluid concrete in bags was deposited. He thus describes the mode of construction :—"During this time, some progress was made at the lighthouse works. The caisson of iron plates to enclose the concrete base had been set up, and about 200 tons of gravel had been deposited upon the reef, where it was exposed to a wash sufficient to remove some of its clayey particles, without carrying it out of reach. As soon as a sufficient quantity of gravel was accumulated, the process of depositing the concrete was commenced. As circumstances did not admit of the usual plan of depositing the concrete in the water in large masses from boxes, the following plan was substituted :—Sheets of tarred canvas were prepared of such

* Minard, p. 58. † Min. Civ. Eng., vol. xxiv. p. 170.

sizes as would fill up the spaces between the piles, and allow two feet round each side to be turned up so as to form large shallow bags. The edges of the tarpaulin were then lashed to wooden rods, which were slung to the piles so as to allow the tarpaulin bag to float slackly on the surface of the water. Two or three hours before low water the work was commenced. The concrete was mixed in the lighters moored alongside the caisson, six measures of gravel being used with one measure of cement and a suitable quantity of water. The materials were thrown into the centre of the canvas bag, which gradually sank to the bottom (generally from one foot to two feet under water) and the bag was spread out evenly over the whole area as it became filled. This was continued until the tide rose nearly to the level of the top of the deposited concrete, when the sides of the tarpaulin were drawn close down over the soft mass, and lashed tight. In this way blocks of from 6 to 14 tons were deposited without the material having been subjected in small quantities to the action of the water. The blocks were generally hard enough on the following day to allow of the exposed parts of the tarpaulin being cut away, and so complete was the set that casts of the cords and the edges of the tarpaulin were often sharply impressed upon the face of the concrete."

Monolithic Structures of Cement, Concrete, or Rubble.—In *Nature* for September 1871, I suggested that lighthouses on rocks in the sea might in certain situations be constructed *wholly* of cement rubble. The advantages of this mode, or of gravel concrete, when used not in separate blocks but in *continuous building*, are the following:—1st, the dispensing with all squaring or dressing of materials. 2d, The suitableness for such works of any stone of hard quality, thus rendering it unnecessary to bring large materials from a distance, or to open quarries for ashlar. 3d, No powerful machinery is needed, as for moving or raising heavy materials. 4th, Saving in the levelling of the rocks for a foundation for the tower. 5th, The ease of landing on exposed rocks small fragments of stone, as compared with the landing of heavy and finely-dressed materials.

A beacon 13 feet high and 10 feet diameter was constructed under Messrs. Stevenson's directions in a very exposed situation near Isle Earraid, Argyleshire, which has stood perfectly for the last two years.

The walls of harbours may also be constructed of continuous building, protected by temporary piling and close planking. The best mode of keeping out the water, as adopted by Mr. Balmer at the Duke of Richmond's harbour of Port Gordon, is to make a saw-draft in each plank, and to place a thin plate of iron between the planks. An example of this continuous or monolithic building is shown in Plate XVIII.

Mixers for Concrete.—There are different kinds of machines by which the concrete is thoroughly mixed, and a saving of cost effected. In that of Mr. Messent the number of revolutions to mix the concrete thoroughly is 12.

Concrete Cylinder Foundation.—Mr. Deas of Glasgow has obligingly furnished me with the subjoined particulars of the founding, by means of concrete cylinders, which he has adopted at the Plantation Quay, Glasgow, and which are shown on Plate XVII.

"The cylinders are made of concrete in sections 2 feet 6 inches deep. The concrete is thrown into wooden moulds placed on a carefully-levelled platform. A section of a group of three 12-feet cylinders measures nearly 17¾ cubic yards, and weighs about 33½ tons. To make the section easier lifted, it is made in segments of three and four pieces alternately, as shown in Plate XVII. : the heaviest of these segments is 11½ tons, and the lightest 9 tons 8 cwts. Mr. Milroy, the contractor for Plantation Quay Extension, makes three complete sections per day of ten hours. Of course this could be increased to any extent by the increasing of the plant. Mr. Milroy has one Blake's stone-breaker, and one Messent's concrete-mixer, constantly at work to produce the above. About three weeks must elapse before the sections can be lifted after being made. The shoes are placed in the trench as near as possible at the level shown on the drawing, and the whole height of cylinders is placed thereon before sinking is commenced. The cylinders are sunk by excavating the material, chiefly sand and gravel, from each well simultaneously, by means of Milroy's patent digger. The average working rate at which each group of three cylinders has been sunk is 2 feet 9 inches per day, including stoppages and delays of every kind. The greatest depth sunk in one day has been 13 feet 6 inches : 8 to 9 feet is a common amount during the first stage of sinking. As the sinking progresses, each group requires to have weights placed on it to force it down, and frequently, by the time the bottom of the shoe reaches 50 feet below cope-level, a weight of nearly 600 tons of cast-iron rings has been stacked on its head. This quantity gives fully 5 cwts. per square foot of total outside surface of each group. To close up the junction of each two groups of cylinders, so as to prevent sand running through from behind, a red pine pole is driven, as shown on the plan, Plate XVII.

"*Cement.*—All the cement used in the work is of the best quality of Portland cement, weighing not less than 115 lbs. per imperial straked bushel, so finely ground as to be able to pass through a sieve of 2500 meshes per square inch, leaving a residue not exceeding 20 per cent, and when mixed up neat in a mould, shall, after seven days' immersion in water, resist without breaking a tensile strain of 350 lbs. per square inch.

"*Concrete.*—The cylinders are of concrete, composed of one part by measure of cement, to five of sand and gravel, or broken whinstone, and the concrete used in the rubble building is of the same strength."

Mr. Messent has obligingly communicated the following notes as to the proportions of concrete :—

"For concrete blocks or ordinary walls, $6\frac{1}{2}$ parts clean gravel or shingle, $2\frac{1}{2}$ parts sand, and 1 part Portland cement.

"The Portland cement should weigh not less than 112 lbs. per striked bushel, lightly filled or sifted into the measure, and, if made into test-bricks, immersed in water as soon as they will hold together, should, after seven days' immersion, require at least $2\frac{1}{2}$ cwt. to break each square inch of the breaking section of the brick.

"The usual area of the breaking section of test-brick is $1\frac{1}{2}'' \times 1\frac{1}{2}'' = 2\frac{1}{4}''$.

"For the above proportions $2\frac{1}{2}$ bushels of cement will be required for a cube yard of concrete; and for the hand concrete-mixer, made to mix $\frac{1}{2}$-yard charges, the cement should be measured into bags containing $1\frac{1}{4}$ bushel—one bag being required for each charge; the hopper being the proper measure for the gravel and sand. Broken slag and broken bricks or granite spauls may be used instead of gravel or shingle.

"For large masses of concrete in foundations or quay walls the concrete may be made with two bushels of cement per cube yard, or one bushel to each charge of hand-mixer; whilst in cases where extra strength is required, over openings, or to resist abrasion, the proportion of cement may be increased to three bushels per cube yard. In each case the concrete may be cheapened without deterioration by placing largish stones in the fresh mixed concrete, care being taken that the stones are all surrounded by and separated from each other by concrete.

"With the hand-mixer a gang of six men, with a boy for attending to the water-cistern, can make from 30 to 40 cubic yards of concrete *blocks*, and a larger quantity of concrete in bulk in a trench."

Mr. B. B. Stoney has been good enough to communicate the following description of his block system, as successfully carried out at Dublin :—

"The blocks are at present used in the lower part of a quay wall, the total height of which is 42′ 10″, say 43 feet. Each block is laid 24 feet below L. W. of equinoctial springs, on a foundation levelled by means of a large diving bell, the chamber of which is 20 feet square, with a tube of wrought iron 3 feet in diameter, rising above high water. An air-cock at the top of this tube admits the workmen in and out, and several men can excavate inside at a time. The water surface inside the bell is quite calm in all weathers, and when the bell is resting on level ground the water is only about half-an-inch deep, so that it gives a plane surface of 400 square feet, which enables the bottom to be levelled with the greatest accuracy and facility. The bell weighs 80½ tons. Each block is 27 feet high, 21′ 4″ wide at base, and 12 feet long in direction of wall, built on terra firma of rubble, set in cement, mortar, and concrete. Each block contains nearly 5000 cubic feet, and weighs 350 tons; and when laid in place 12 lineal feet of the wall is finished at once up to ordinary low water level. No cofferdam, staging, or pumping is required. The superstructure is built in the ordinary method by tide-work and is faced with granite ashlars for the ships to lie against. The blocks are built on land, and after 10 weeks drying are lifted by a floating shears, the barge or pontoon of which is 130 feet long and 48 feet wide, and of this 130 feet 30 form a tank at the aft end; and, when filled with water, this balances the weight of the block hanging from the shears at the other end. A block can be raised with the flood tide, and is generally set the following low water. The rising of the tide is not necessary, as the lift could be made in water with a constant level, but you can easily understand that a rapidly falling tide would not be desirable when lifting a block, as the floating shears sinks some feet under the weight of block and water, say 700 tons, and this, combined with a falling tide, would require the block to be lifted high up in the air. The blocks are laid touching each other, and the range is wonderfully accurate. A recess (vertical) is left in the side of each block, and into this vertical groove a lot of concrete is shot when the blocks are in place; this forms a key, and effectually stops up the little space (about half-an-inch) between block and block. Several hundred feet of wall have now been built, and the work is progressing very satisfactorily."

Iron Concrete.—Mr. Leslie has introduced at the Stranraer pier, which is constructed of timber, a concrete consisting of gravel and iron borings, which seems to have answered its purpose very well. Mr. John Howkins junior, who was the resident engineer, says (in a letter to Mr. Leslie, who has kindly communicated it to me)—

"The quantity of iron borings mixed with the hearting was 160 tons; and, taking the weight of gravel at 20 cubic feet to the ton, the proportion of borings to gravel, in weight, is 1 ton of the former to 17 of the latter, and the proportion in bulk 1 to 34,—assuming 10 cubic feet of iron borings to weigh 1 ton (which, however, I have not the means of ascertaining at present). The borings below low water were all thrown in by a person employed for that purpose, whose duty it was to scatter with a shovel a quantity as nearly proportionate as possible to the quantity of gravel which had been deposited at the side of the piling from the waggons. For the purpose of establishing that proportion, he had a box which was made capable of holding a quantity proportionate to the contents of a waggon. Above low water the gravel was thrown to the sides in thin layers, and the borings added alternately. In digging down at the end of the pier, where the gravel and borings had been acted upon by the sea for two and a half months only, I found the layers of borings caked in a hard mass, and particles of gravel and sand adhering to the sides of the layers; showing that the concretionary influence of the borings is extending, and will, in all probability, in course of time completely bind the intervening layers of gravel. I had also an opening made at the back of the slip coping, which has been filled in five or six months; and there the gravel, which is much coarser than at any other part (the sand and finer particles having been washed down and away by the sea), was very strongly coloured, and showed much the same appearance as at the end of the pier, only in a more decided degree. With regard to the coating of the timber, there is a decidedly rusty appearance on the outside of the piling, immediately above low-water mark, quite observable at a considerable distance; and, on a closer examination, the joints of the sheet piles are, in a number of places, seen to be giving forth that slimy discharge similar to what is so commonly observed in the neighbourhood of ironstone."

Asphaltic Masonry and Concrete.—I have tried, at the island of Inchkeith, some experimental masonry, which is cemented together with British asphalte. At the same time, the experiment was successfully tried of letting down under the surface of low water stones and hot asphalte, placed in canvas bags, which were pressed down upon the irregular rocky bottom, so as to equalise it, and render its surface ready for founding on. This substitute for mortar or cement in rubble and ashlar work seems capable of resisting the chemical action of salt water, for at Inchkeith it has stood for several years. Mr. Manley mentions that asphalte was proposed for harbour works in France, but that the risk of its decay prevented its adoption. He does not mention, however, whether any experiments were made in order to test its qualities.

Carbonite Cement.—While the proof-sheets of this book are passing through my hands, my attention has been called by Mr. H. C. Paterson, of Glasgow, to a new material for building and concrete, which plainly presents features of marked interest, the invention of Dr. George Hand Smith, of New York, a prominent American metallurgist. It is the result of a combination mainly of hydrocarbon of varying densities with clay, chalk, or gypsum, whereby concrete and all kinds of material for building purposes can be produced. The constituents are stated to be cheaper than those at present in use, as well as of a more durable character, being unaffected by frost or exposure, while they resist the action of the most powerful muriatic, sulphuric, and hydrochloric acids. It is further stated that the concrete can be made of any required density.

The following extracts from Dr. Smith's statement regarding this new material will be of interest to the engineer. One of the great recommendations of this cement is the fact of its being fit for use immediately after being manufactured. There is also claimed the further advantage that the blocks are capable of being fixed to each other under water by an indestructible cement of great binding power, which is also prepared in the process of manufacturing the carbonite. The first process by absorption is thus described :—

"The first step in the process is the proper preparation of the vitally important carbon bath, which consists of a suitable tank or tanks to receive the chalk, etc., containing in a liquid state, insoluble, non-volatile or free carbon, suspended and combined with volatile hydrocarbons in the definite proportions desired. Other chemical substances are also added in moderate quantity, as experience dictates, to modify the action in view—such, for instance, as alumina, iron, etc.—although the relative action of the two forms of carbon is an ever-constant and essential feature. The volatile carbon becomes the vehicle for inducting the free carbon into the chalk, etc., conveying in their turn the other elements associated or combined with them. The volatile carbon, on removal, evaporates or is driven off, leaving the free carbon permanently fixed. The proportions of the carbon are determined by the hydrometer, easily insuring accuracy of result. The bath is kept hot by steam, or a grate. For some purposes it may be cold. It should always be quite liquid. Whatever chemical agents are employed to modify the action of the carbon bath, neither in quality or quantity do they unfavourably affect the question of profit.

"Into this bath, by a simple apparatus of an open iron bucket or crib, controlled by pulleys, a large amount of the clay, chalk, or plaster, first moulded or cut into the desired form, or in masses or blocks to be afterwards turned or cut into shape, is at once immersed, whereby it becomes rapidly charged with the carbonaceous mixture. By means of successive cribs, which are re-loaded and discharged in regular turn, the impregnation is rapidly effected. For some products the volatile hydrocarbon, etc., is driven off by the cooling of the same on removal from the bath; for others it is accomplished in a different manner, as next described."

The second process is by compression, of which Dr. Smith says—

"In the manufacture of bricks, large blocks, waterproof for basements, docks, etc., where the shapes are plain and uniform, it is more economical to effect the moulding of the brick, drain pipes, etc., and the carbonisation, at one and the same time. Machines for this purpose are fully perfected, that will turn out from 5000 to 8000 bricks per day, and drain pipes in proportion.

"The carbonisation is effected by combining the raw or common clay with the carbonaceous solution in the ordinary hot mixer; thereby rapid action ensues, the confined vapours playing a part, the clay becoming more or less anhydrous, or giving up its combined water; its place being supplied by the elements from the solution. The excess of vapour may be saved and utilised as before mentioned. These carbonised products of clay or chalk, on cooling, would become hard; are immediately compressed into shape, becoming much harder by their condensation. They have afterwards simply to cool, when they are ready for use."

Dr. Smith's other process by compression is the following :—

"In this method, instead of combining the earthy constituents and the carbonaceous solution, while hot, they are simply mixed together cold, or nearly so, or treated by immersion, and then compressed into the forms required. They are afterwards allowed to remain for a short time in the drying chamber, the same as described above, whereby the same change occurs; the volatile carbon being saved for further use, while the solid elements are permanently fixed. Chalk in pieces for cutting, or in slabs for marble, may be thus treated without compression.

"The advantage of this method consists in the ability to modify the *quality* of the products, where it is desirable to retain a larger proportion of all elements contained in the solution, as in making some kinds of brick and blocks, artificial marble from chalk, etc.

"In addition to bricks, waterproof blocks of the very best description may be prepared, of large size, from five to ten tons, without the use of power-presses, suitable for construction of docks, piers, embankments, reservoirs, tunnels, waterworks, cellars, and all subterranean works.

"The remarkable ability of a product that has never been burned, to resist high pressure, is shown by the experiments of Mr. Kirkaldy, one of the highest authorities, whose Report is appended, where samples indicated between 7000 and 8000 pounds to the square inch as the crushing test, or over 61 tons on a four-inch cube, or over 500 tons to the square foot.

"The following Table (Molesworth's Formulæ) gives comparative figures :—

Materials.	Pressure per square inch.
Portland Cement	1000 lbs.
Common Brick	1500 ,,
Portland Stone	3700 ,,
Sandstone	5000 ,,
Granite	8000 ,,
Clay Carbonite	8000 ,,

"Repeated tests of my own have since ranged as high as 10,000 pounds to the square inch, according to the proportions and method employed. 8000 pounds may, therefore, be safely regarded as a fair average."

"Report of Tests made by Mr. David Kirkaldy on Carbonised Clay.

"'Results of Experiments to ascertain the resistance to a gradually increasing Thrusting Stress, of five cubes, received from Dr. Smith:—

Test No.	Description.	Dimensions. Inches.		Base Area. Square Inches.	Stress in pounds when	
					Cracked Slightly.	Crushed. Steelyard dropped.
H						
1502	Material	3.97	4.08 × 4.08	16.64	112,370	123,870
1504	Do.	3.58	4.09 × 3.98	16.27	105,840	115,860
1503	Do.	3.85	4.00 × 4.02	16.08	89,280	109,640
1501	Do.	4.12	4.04 × 4.10	16.56	99,840	104,980
1500	Do.	3.88	4.08 × 4.00	16.32	89,520	93,560
...	Mean	......		16.38	99,370	109,582
...	lbs. per square inch	......		...	6.067	6690
...	Tons per square foot	......		...	390.2	430.2

"'All bedded between pieces of pine three-eighths inch thick.'"

Measurement of proportions of the materials for Concrete.—Mr. J. F. Bateman has long employed the following mode of determining the relative proportions of the various kinds of material. The method was to take "any vessel, no matter of what dimensions, so that they were correctly known, and to fill it with as much gravel as it could be made to hold by shaking or beating it down—if gravel and sand were to be mixed, then by putting in afterwards as much sand as the vessel would hold in addition, and shaking that down amongst the gravel, the quantity of gravel and of sand being respectively measured as they were put into the vessel. When as much sand as possible had been shaken down, as much water was to be poured in as the vessel would hold; the quantity of water would then represent the lime required, the theory being that each particle of sand or gravel should be imbedded in or surrounded by a matrix of cement; and if the amalgamation of the materials were perfect, then the water which undoubtedly surrounded every particle would correctly represent the lime or cement to be used; but as such perfect amalgamation could not be expected, a somewhat larger quantity of cement than water was employed. By this rule a thoroughly cemented mass of concrete was obtained."*

Mr. Bremner's Pontoons.—The bold project of the late Mr. Bremner, of Wick, for putting in the foundations of low-water piers, merits notice. Mr. Bremner proposed to construct, in some adjoining place of shelter, enormous *pontoons* of timber, in which the under parts of the work were to be built, and afterwards floated, in favourable weather, to the desired spot, and carefully grounded. Such a plan might perhaps be found economical and suitable in some situations; but the great difficulty would be to fit the bottom of the pontoon to the irregularities of the ground on which it was to rest.

Mr. Rendel's Method of Depositing the Pierres Perdues.—The late Mr. Rendel introduced the improved and very valuable method of depositing the *pierres perdues* or rubble, which is now generally used in the construction of large breakwaters; this method he first employed at Millbay Pier, near Plymouth, in 1838, in a depth of 38 feet; and afterwards, on a still larger scale, in the construction of the breakwaters at Holyhead and Portland. The improvement consists in depositing the rough materials from stagings of timber, elevated a considerable height above high water. The stones are brought

* Min. Civ. Eng., vol. xxxvi. p. 242.

on the staging in waggons, through the bottoms of which they are discharged into the sea. The principle on which these stagings are designed is that of offering the smallest possible resistance to the sea, the under structure consisting of nothing more than single upright piles for supporting each roadway.

The late Mr. Rendel kindly communicated to me the following description of his staging, in a letter which is still worthy of preservation as coming from the original proposer:—

"I use no timber braces of any kind, as these offer more resistance to the sea than strength to the staging. At Portland, however, where any accident would be a serious evil, owing to our employing convicts in the quarries, we stay the piles with iron guys, fixed to Mitchell's screw moorings, and also truss the outer piles in each row with iron rods. We also fix the piles in the ground with a screw. At Holyhead, however, we only attach to each pile boxes filled with small stones, for the purpose of getting them into a vertical position, and use no stays or guys of any kind. The superstructure consists simply of balks of timber, with rails laid on them to carry the waggons. The piles are placed in rows 30 feet apart, and the ease and certainty with which the staging is constructed is such that a length of 30 feet, including the screwing in of the piles, the laying down of the roadways, and all minor works necessary to make them fit to carry the waggons, never occupies more than one working day and a half, and often less. The length of the piles that we are now using varies from 84 to 90 feet, the depth of water at both Holyhead and Portland being about 11 fathoms.

"Of the strength of the stage you may judge from its carrying on each roadway as much as three waggons, weighing in the gross twelve tons each.

"The advantages of the staging are obvious. It contributes greatly to the consolidation of the stone; it makes a greater length of breakwater to be under construction at the same time; and it enables the deposits to be carried on without interruption, almost in the heaviest weather. As an instance of this, I may remark that my resident at Portland informs me that the waggons and locomotives were engaged yesterday at a time when such a sea was running that large bodies of spray were thrown 55 feet above the water-level. As a proof of the facilities which the stage affords for rapidity of construction, I should state that we have deposited this year at Holyhead, where free labour is employed, nearly one million tons of stones. The loss from accidents to the stage is comparatively small on its first cost, and when spread over the cost of the whole works it is a mere trifle."

Greenheart Timber Staging.—At Pulteneytown harbour works, as already stated, it was found that the piles of the staging for depositing the rubble required to be of greenheart timber; but even these were broken in large numbers. The piles were invariably broken by the waves at about the level of high water. It will be seen from the accompanying table, which is taken from a paper by Mr. Justen in the *Builder*,* that greenheart, irrespective of its valuable property of resisting to a great extent the worm, possesses such superior strength and specific gravity as to render it by far the best material that can be employed for such a purpose.

TABLE showing Specific Gravity of Different Timbers.

Timber	Specific Gravity
Ironwood	1.210
Greenheart	1.200
Sabacue Wood	1.100
Brazil Wood	1.100
Oak	.885
Beech	.852
Ash	.854
Elm	.800
Fir	.657
Cedar	.561

* Tables by Joseph Justen, on the properties of timber, republished in *Engineering Facts and Figures*, by A. B. Brown. Lond. and Edin. 1864, pp. 359, 364.

TABLE of Strength and Weight of Different Timbers.

	Constant.	Weight per Cubic Foot in lbs.
Yellow Pine . .	358.5	25.687
Baltic Pine . .	444.	29.062
Red Pine . .	467.	33.437
Ash . . .	517.75	41.812
English Elm . .	592.25	37.312
Pitch Pine . .	629.00	45.750
American Elm . .	631.50	45.312
American Oak . .	653.50	44.875
African Teak . .	673.50	60.562
Mora . . .	691.00	71.250
Sabacue . . .	854.25	59.687
African Oak . .	869.50	...
Greenheart . .	1079.50	69.750
English Oak . .		53.312
Indian Teak . .		38.125
Ironwood . .		73.500
English Larch . .		32.562

Iron Lattice-Work.—I have found reticulated framings, consisting of galvanised malleable iron bars, useful for protecting the masonry of piers above low water, where they can be *paid* over with hot pitch from time to time as required. The bars, which require very few attachments to the masonry, can be welded or rivetted together, so as to form large frames, having open meshes of from one square foot upwards; and if made with ring joints would form a sort of *chain net-work*. Iron lattice-work has been used in protecting the sloping surface of a weir, in a river which is subject to sudden floods, as also in a sea-wall at Arbroath. The principle of the iron lattice-work is the combination of a large part of the strength which is due to a whole plate of metal, with a greatly reduced surface, thus giving an abundant outlet for the escape of condensed air.

CHAPTER XI.

ON THE EFFICACY OF TIDE AND FRESH WATER IN PRESERVING THE OUTFALL OF HARBOURS AND RIVERS.

Pools and Shoals due to variations in the Flood-water Sectional Areas—Flood-water and Summer-water Sectional Areas inversely proportional to each other—Gravelly Rivers—Contraction of Estuaries—Relative Values of Salt and Fresh Water for Scouring—Value of Tide Water on Banks—Abstraction of Fresh Water—Symmetrical Section—Effects of Embankments.

THE commercial value of our harbours and navigable rivers is principally dependant on the depth of the channels which connect them with the ocean. It seems, therefore, scarcely possible to over-estimate the importance of ascertaining what is the principal agent that scours and preserves the depth of such channels, and what are the conditions under which that agent operates to the greatest advantage. Yet it is one of those subjects on which much difference of opinion prevails, and there is consequently a corresponding want of agreement among engineers as to the principles which should regulate the design of works for improving tidal channels. The engineering of rivers and estuaries has been very fully treated in Mr. D. Stevenson's "Canal and River Engineering." The only part of the subject which requires notice here, as more especially belonging to harbours, is the relative values of salt and fresh water as scouring agents, and I have to refer the reader for further information on this and all other subjects relating to River Engineering to Mr. Stevenson's Treatise.

Pools and Shoals in Rivers above the influence of the Tide due respectively to Contractions and Expansions in the Flood-water Sectional Areas.—It is unquestionably a fundamental law of hydrodynamics that by contracting a channel you secure a deeper track if the soil be so soft as to be removable by the increased velocity of the current.

I had lately occasion, with reference to a question of law, to inquire into the causes of the formation of certain pools and fords which existed in the upper parts of a river to which the tide had no access. The conclusion at which I arrived was that, where not due to differences in the nature of the soil, shallows were occasioned either by bends in the river, in which case they are formed on the convex side of the stream, or to enlargements of the flood-water channel; so that, where the summer water area is small and the velocity great, the flood-water area will be large and the velocity small, and *vice versa.* Conversely, that deep pools were due either to sudden bends, in which case they are formed on the concave side of the stream, or to sudden contractions of the flood-water sectional area, owing to the banks above the summer water level being high and steep. In other words, *where the stream has a straight course, and the soil is homogeneous, there is a constant relation subsisting between the flood-water and summer-water sectional areas, and these two are inversely proportional to each other.*

In Gravelly Rivers the great leading features of Depth and Direction are due to the occasional action of very heavy Floods, and not to the river in its ordinary state.—In rivers which pass heavy gravel, the bed, both as regards depth and direction, will depend upon great land floods, which are alone able to scoop out to a great depth heavy gravel and boulders. I believe that the river Spey in Morayshire still bears marked traces of the great Morayshire floods in 1829, when the discharge was greatest, or rather when the ratio subsisting between the sectional areas and the discharges was the greatest, and when the scour at the bottom would of course vary with those ratios.

The greatest floods should therefore make the greatest excavations at some places and the greatest deposits at others, just as the relative magnitude of discharge and areas shall determine. If a straight river were of regular section, or rather of sections everywhere proportional to the discharge, then the floods would deepen the channel uniformly throughout, without the formation of pools and shoals.

The pools formed by the greatest flood will not remain of the same depth, for the river, under the influence of smaller floods, passes periodically certain sizes of gravel, washed down by rain from the land, in quantities inversely proportional to the cubes of the diameters of the gravel—to the density of the materials—and to the duration and amount of the discharge of water. The first considerable flood after the greatest, will not move the heaviest class of boulders which the greatest flood had been only able to move a very short distance ; but gravel of lesser size will be carried down the steep slope at the upper end of pools. When this gravel came into the pool it would be in deep water, and resting on a comparatively level bottom, and not so likely, in such circumstances, to move, as during greater floods. The tendency of a succession of such floods is obviously, though very slowly, to restore the level of the pool to what it was, before the occurrence of the greatest flood.

But these remarks apply only to what has been termed the "river proper," to which the tide has no access. In such a case no changes in the alveus of the stream can possibly affect the amount of the scouring agent, which consists of rain from the uplands, and the produce of springs ; and which amounts, less absorption and evaporation, *must* be passed through every section of the channel on its way to the sea.

Contraction in an Estuary may reduce amount of scouring power.—But the case is different with a tidal river or branch of the sea, in which, although a local contraction will undoubtedly produce an increased *local* depth, it does not follow that such contraction may not injuriously reduce the depth nearer the sea ; for if water be excluded from the banks of the upper estuary, there is a risk that it may cease any longer to enter the navigation.

Relative Values of Salt and Fresh Water as Scouring Agents. —Some engineers (among whom was the late Mr. Robinson Palmer) have gone so far as to consider the fresh water the only efficient scouring power, while they regarded the tide water as rather an evil. In April 1812 the late Mr. Robert Stevenson discovered in the river Dee at Aberdeen harbour, that when the salt water of the ocean enters even small estuaries during flood-tide, it does not, as had before been supposed, oppose and commingle, in ordinary states of the weather, with the outgoing fresh water; but from its superior density, as proved by the hydrometer, insinuates itself along the bottom of the channel, and raises the fresh water above it, which still continues its regular discharge outwards into the sea, in a film separate from the salt water. In 1817 the late Professor Fleming, in repeating at the Tay similar observations to those made at Aberdeen, found corroboration of Mr. Stevenson's result, and that, 18 miles from the river's mouth, the specific gravity was greatest at the bottom. He farther found that the marine vegetation *adhering to the bottom, though not appearing at higher levels,* also bore its testimony to the constancy of the operation of this law. He noticed that at Flisk, "and even farther up the river, the *Fucus vesiculosus* (the species commonly cut for making kelp) not only vegetates, but in its season appears in fructification. But that which proves in a still more decisive manner the action of the inferior stratum of salt water at the place, is the growth of the coralline termed *Tubularia ramosa*, and another of a different genus, closely resembling the *Sertularia gelatinosa* of Pallas."

Tidal Guts.—Perhaps as obvious a proof as can be adduced of the independent and separate action of the flood-tide, is the formation in the lower parts of estuaries of what may be called *tidal guts*. These are subsidiary or lateral low-water tracks, having always their greatest width next the mouth of the river, and contracting gradually upwards. In many instances, as for example in the Tay and in Lough Foyle, these tidal guts, though often rivalling, and even exceeding, the depth of the main channel through which the fresh water passes, are *blind* channels, ending in a *cul de sac* at their upper ends. No stronger proof than this need be adduced of the great power of the flood-tide in excavating and moulding for itself those deep channels which are altogether separate from that which is occupied by the fresh water.

Allegation that the greater Velocity of the Flood Tide which proceeds landwards must cause a Silting-up of the Estuary.—Mr. Palmer stated that "the effect of the flowing tide in raising the bed of the river exceeds that of the ebbing tides, and hence we may conclude that the depth of the channel is entirely and exclusively dependent upon the water which is derived from the uplands." The question may certainly very properly be asked how it is that in any case where the flood tide has a greater velocity than the ebb, it does not bring in more sand than the ebb takes out. At first sight one would certainly be led to infer that, although the flood tide no doubt excavates deep tracks in the lower reaches of navigations, yet in the upper parts of salt water *bays* into which no rivers discharge, and where there are no land freshes, it should cause accumulations. The only answer that I can think of is the fact that the flood tide, in bringing up sand from the ocean, has to raise it from a lower to a higher level, or, in other words, has to work against gravity; while the ebb, in dragging the particles from a higher to a lower level, is assisted by gravity.

It is therefore quite in accordance with mechanical principles that the ebb, though having a less velocity, should, in pushing the particles down hill to the ocean, operate as much on the bed, as the flood tide with a greater velocity in pushing the particles up hill from the ocean. In estuaries where there is a large discharge of fresh water, the case is clear enough, because in these the ebb is assisted by the outgoing fresh water, which also, in some states of the weather, largely interferes with the translating power landward of the flood tide.

From these premises we may conclude that both the fresh and tide waters are useful in preserving the navigable depth. Were it otherwise we should find the very opposite of what is seen everywhere in nature. The lower down we go in an estuary, we generally find the larger low-water sectional area; whereas, if the fresh water be the only scouring power, and the tide water an evil, we should find the greatest depth and the greatest sectional area in the "river proper."

Value of the Tide Water which covers the Banks or fills the side Creeks of a Navigation.—Some engineers, who do not dispute the efficacy of the tide water as a scouring power, are not prepared to admit that the water, which at high tide occupies the banks and side creeks of a navigation, is of much, or perhaps of any value. The difference of opinion which exists among engineers seems to arise from the want of any method of determining, by direct observations, whether or not such tidal water has in any case operated effectually as a scouring agent. The daily varying amounts of tidal water which are propelled into an estuary from the ocean; the ever-changing discharges of the land-waters, being sometimes very small and at others prodigiously increased in volume; and the heterogeneous nature of the materials forming the bed, which at different parts of the same river consists of gravel, sand, or mud, exhibiting an endless variety in the sizes and coherence of the particles—present an almost hopeless complexity for the mind to grapple with, or for even elaborate observations to unravel.

From a comparison made some years ago of different low-water sectional areas of river estuaries and of open bays in their state of nature, I remarked in the first edition of this book, that in many estuaries and creeks *the low-water sectional areas seemed to increase directly as the quantities of tide water that passed landwards of such section lines.* By thus comparing the sectional area at any point, with the area at a point a little farther down the estuary, we free the question from the difficulty of dealing with the unknown action of the freshes, which is nearly the same at both places, as well as of the ever-varying amount of the tides, which is also the same at each. We should therefore, where the bottom is of uniform consistency, find a progressive increase in the *low-water* sectional area, proportional to the progressive increase in the amount of tidal water as we approach the sea, because the amount of tidal water is always increasing, while the land-freshes in most rivers remain nearly the same, at least for short distances. I do not, however, venture to assert that the principle has been sufficiently established to admit of general application, nor do I believe that it can; but I regard it as a convenient method of illustrating the beneficial influence which, in some estuaries at least, is exerted by the water which covers the banks at high tide.

Where a a' a'' are low-water sectional areas of the channel in approaching the sea, and $c\ c'$ are the intermediate *high-*water capacities, then

$$a'' = \frac{c'}{c}(a' - a) + a'$$

The following are examples taken from the Tay below Newburgh, and from Belfast Lough :—

River Tay.

Actual Capacities.	Capacities by Calculation.
87,355	92,000
109,646	99,600
104,226	155,652
Mean 100,409	Mean 115,751

Belfast Lough.

Actual Capacities.	Computed Capacities.
242,651	243,000
318,208	304,000
367,913	319,000
333,721	387,000
453,623	463,000
Mean 343,223	Mean 343,200

In the narrow artificial channel of the Dee, Cheshire, the efficacy of a given quantity of tidal water was, as might have been expected, greater than in navigations which were left more nearly in a state of nature.

It must, therefore, be kept in view that what has been stated is merely of a general character, and is not to be regarded as of constant local application, for the value of a cubic yard of water depends on its level in the estuary. The following remarks by Mr. D. Stevenson should therefore be kept in view in dealing with individual cases.*

* *Canal and River Engineering.* A. & C. Black, Edinburgh, 1872.

"It will readily be seen that the efficiency as a scour of a cubic yard of water filled and emptied by *every tide*, as compared with that of a cubic yard filled only *five times during every set of spring-tides*, is in the ratio of 730 to 144, not to mention the more effective scouring power of water discharged after half-ebb, as compared to a similar quantity discharged, for example, during the first hour after high water.

"The value of the water as a scour is therefore influenced both by its *volume* and by its *level*, and may be expressed as follows :—

$$S \propto V\ T,$$

where V = the volume or cubic feet of water space above the low-water level of the estuary.

T = the number of times it is filled by the tide throughout the year.

S = the effective scouring power.

"The only other consideration that should be kept in view is that of two spaces, V, V′, of equal capacity, and filled *every tide*, that which is lowest in position will be most effective in operating on the low-water channel. These values must of course be held applicable only to different conditions of the *same river* where the hardness of the bottom to be scoured and other circumstances remain unaltered."

EFFICIENCY OF FRESH OR RIVER WATER AS A SCOURING AGENT, AND THE EVILS OF ABSTRACTING IT FROM A NAVIGATION WHERE THE BOTTOM IS SOFT.

Abstraction of fresh water where the soil is soft will cause reduction of capacity of alveus and diminution of the quantity of water that enters from the ocean.—The evils of abstracting fresh or river water from a navigation are twofold. *First,*

Directly, by the loss of this constant scouring power. *Second,* Indirectly, by the diminution of capacity of the alveus, which is produced by the rising of the bottom and sides consequent on the reduction of the fresh-water scour, so that the difference in amount of tidal water due to the decreased area of the alveus is excluded from the navigation. It is obvious that the contraction of the alveus must reduce the quantity of water that used to come in from the sea; for if we suppose that a stream is diverted from its natural outfall, and allowed to discharge at another part of the shore, the sectional area of the stream at any given place between high and low water will, if the shore be rocky, vary inversely with the fall of the beach at that place. But if the bottom be soft mud, the tendency of the stream will be to equalise its gradient by cutting (if the soil be homogeneous) a straight channel with a uniform gradient over the beach, due to the total distance between the high and low water margins divided by the rise of the tide. It is obvious, therefore, that after a sufficient time has elapsed, the stream will have cut an alveus in the mud of much larger sectional area than the sectional area occupied by its own *fresh* water. Nothing is more common than alvei, such as I have described, having at low water only a very small area occupied by the effluent fresh water.

Now, as the whole alveus is filled at high water, it follows that the comparatively trifling stream has secured an additional scouring agent from the sea, very much greater than is represented by the fresh-water sectional area multiplied by the distance between the high and low water margins, so that with the ebb-tide there is available for scouring purposes the constant fresh-water discharge, plus the whole salt-water contents of the alveus.

Even in narrow artificial cuts, such as the Dee at Chester,

where the flood-tide comes in with a bore and the fresh-water current is reversed and sent landwards, the whole amount of fresh water impounded during the time of flood tide is altogether insufficient to fill the estuary. Mr. D. Stevenson found the high-water capacity above Connah's quay to be about 220 millions of cubic feet, while the whole aggregate ordinary discharge of fresh water during flood, amounts to only 3 or 4 millions of cubic feet. Hence we see that the fresh water in this case is vastly more efficacious indirectly than directly.

Abstraction of fresh water, though it causes a reduction in capacity of alveus, does not necessarily reduce the depth in the channel.—It does not follow, however, that the *bottom* will in all cases be raised by the abstraction of fresh water, although the capacity of the alveus is diminished. The bottom, for example, may have consisted of gravel of so large a size as to have defied the eroding action of the original current, and therefore there may exist a gradient sufficient to generate a current which precludes the deposit of such light materials as the river brings down. If this be the case, instead of the *depth* being affected, it will only be the *breadth* that is changed. The reduction of the discharge will, therefore, alter the relation of breadth to depth, without decreasing the depth.

Symmetrical Section.—Had the soil in the case which we have supposed been homogeneous—that is, had the soil of the bottom instead of being heavy gravel been the same as that of the sides—the symmetry of the section would have been preserved, and such a section may therefore be termed the *symmetrical section.*

Forced or Artificial Section.—The symmetrical may, however, not be the most suitable for the navigation, as it may be better to increase the depth at the expense of the breadth. But then, to reduce the scouring agent is not the proper way to effect this object, but the very reverse, for that would reduce the section without increasing the depth. If a *forced section* has to be substituted for the symmetrical section, the proper expedient is to erect stone walls at the sides, so as to destroy the homogeneity of the material forming the perimeter of the channel, and thus to alter the relation of depth to breadth.

Abstraction of a portion of the fresh water which at present is capable of disturbing and rearranging the constituents of any kind of soil, may occasion a greater deposit of the same, or of a finer kind of material, but not of a coarser.—If what forms the existing bottom be the same as has been brought down by the present scouring agent, then any diminution of that agent will cause a new deposit of the same kind of material, or of a finer kind, but not of a coarser. This is clear, for if the original amount of scour be insufficient to lower any further the present level of the bottom, it will be insufficient to move a heavier or coarser material. *If the reduced current be still able to move the materials stated in the first column of the Table,* then the deposit shown in the second column will take place; and if the current be still more reduced, than the order of succession in the third column becomes possible.

TABLE of Order of Deposit of Materials when Backwater is reduced.

Nature of Present Uppermost Deposits.	Fresh Deposits that are certain.	Fresh Deposits that are possible.
If silt . .	Silt	
,, mud	Mud	Silt.
,, sand	Sand	Mud or silt.
,, gravel . . .	Gravel	Sand, mud, or silt.
,, boulders . .	Boulders . . .	Gravel or sand, or mud or silt.

If the whole summer water discharge of a land stream be diverted from its proper outfall for manufacturing or other purposes, a deposit, unless the water were absolutely clear, must take place throughout the whole navigation. But where the upper reaches are very narrow, and thus have a small flood-water sectional area, the deposit that will take place during summer droughts will be only temporary. Whenever the heavy winter freshes, which cannot be stored or diverted, descend from the uplands, the deposit that took place during the droughts in those places where the flood-water sectional area is small, will, I believe, in most cases be removed, and the original area restored.

In the lower reaches, or at other parts where the flood-water sectional area is large, the result will be different. At the period of slack water of the tide perfect stagnation must exist, and a greater deposit will take place on the flat banks than there would had there been even a small downward current, to keep the water near the middle of the channel in motion, and to generate side currents, which would be found useful in reducing the tendency to deposit. After the tide has passed out, and the deposit on the upper parts of the flats has been subjected to the sun's rays and to the action of drying winds, it may attain a coherence so great as to resist the next winter floods, which act most powerfully in the *filum fluminis*—and not close up to the margin of the marsh land where this deposit has taken place.

Beneficial Action of the Waves on the Lower Reaches.—The permanent accumulation to which we have just referred will be greatest where there are projections of the land to afford shelter; but if waves get access from the ocean, or if the estuary be in itself wide enough to admit of the generation of waves of a foot or two in height, the deposit will be to a large extent broken up and removed during high winds.

Alleged Increase of Tidal Water by contracting the Alveus.

—Some engineers have stated that if the channel be gradually contracted so as to form a trumpet-mouthed river, like the Wye at Chepstow, the tide will, on the principle of the conservation of forces, attain a higher level in the upper reaches of the river, so that the quantity which enters the navigation will not really be diminished. To this principle no objection can be offered, for we find in the British Channel that at

Swansea, a spring tide rises		-	30 feet.
At mouth of Avon	,, -	-	40 ,,
At New Passage	,, -	-	50 ,,
At mouth of Wye	,, -	-	60 ,,
At Chepstow	,, -	-	70 ,,

But the practical question is whether it be possible by any artificial works to cause a rise which will be equivalent to the water excluded by embanking. When we have an initial tidal column of 60 feet, as at the mouth of the Wye, it may be possible by artificial works to cause a very considerable rise in the height of the tide ; but with a column of 16 or 18 feet, we cannot expect that any very appreciable effect would be produced unless by works of unusual magnitude. Indeed, almost the only case with a small rise that I know of, is that of the Tyne, where, as appears from Captain Calver's report, there is a rise of about six inches additional at Newcastle, due to operations on the most extended scale. But it must be observed that this improvement has not been effected by contraction, but by dredging out the bottom.

Facts regarding the Decrease of Depth due to the Embankment of Estuaries.—At the port of Ostend the evils of embanking were clearly shown. Minard states that in 1626 an embankment was made which excluded the tide from a large portion of submerged land. The consequence was that the harbour began to silt up till there was a depth in the channel of only 1 metre. In 1662 a cut was made in the embankment so as again to admit the tide, and in 1698 the channel was found to be from 13 to 17 metres in depth. The cut was again closed in 1700. In 1701 the depth was still good, but in 1716 complaints began to be made that the channel was silting up. In 1720 the channel was almost closed, when new cuts were made and considerable deepening resulted, but the back lands were by this time so much raised that some of the cuttings were shut up as useless. Further silting went rapidly on till 1810, when the depth was reduced to 2 metres. The mode of inundation was then abandoned, and artificial scouring by sluices was substituted with very indifferent success.

The river Dee, in Cheshire, also affords important data on the same subject. At Parkgate, in 1664, the depth at low water was 15 feet, and in 1732, before the river was embanked and diverted, there was still a depth of 15 feet at low water and 39 feet at high water. The River Dee Company was incorporated in that year for the purpose of diverting the channel to the south side and embanking the estuary. The channel at Parkgate is now quite dry at low water, and has a depth of 10 feet at high water ; and near Flint, on the south side, where the navigable channel now exists, the depth at low water is 3 feet, and at high water 23 feet, giving a rise of 20 feet. Parkgate is 4½ miles below the end of the artificial causeway formed by the River Dee Company, but on the opposite side, and the point near Flint selected for comparison is on the same side of the estuary. As the depth near Flint at high water is 23 feet, and that at Parkgate 10 this gives 13 *feet of silting, which is due to diverting the channel.* But there was formerly a depth of 39 feet at high water at Parkgate, instead of as at present 10, which gives 29 feet of silting due to diversion and embanking. The results then, stand thus :—

13 feet of silting due to diversion only.
29 „ „ due to diversion and embanking.

Leaving 16 „ „ *due to embanking only.*

Conclusion.—On the whole, then, it appears that, unless in peculiar cases, embankments of any considerable extent within the tidal compartment of navigations and estuaries ought to be resisted, as well as all abstraction of the upland waters for manufacturing or other purposes. The mode of concentrating the action of the currents by high walls is therefore to be avoided, especially as the same result can be effected, without objection, by means of low training-walls, which, by guiding the first of the flood and the last of the ebb, fix most effectually the *filum fluminis* in one given direction, and thus concentrate the scour without necessarily excluding tidal water from the estuary.

CHAPTER XII.

MISCELLANEOUS SUBJECTS RELATING TO HARBOURS.

Movement of Shingle—Shoaling of Harbours—Artificial Scouring—Dredging Sand-drifts — Danger of deepening Harbours — Commercial Value of Channels proportional to Cubes of their Depths—Draughts of Ships—Scend of Waves—Ruling Depth of Harbours—Lighthouse Apparatus—Timber Ponds—Harbours with two Entrances—Screw Piles—Economy of Time—Hydraulic Apparatus—Coal Staiths—Steam Cranes—Tonnage of Ports—Government Aid.

Movement of heavy Shingle by the Surf.—Wherever the heaviest waves strike obliquely on the shore, the shingle, if there be any, travels across the beach, and is very apt to fill up the entrances to harbours. Examples of this may be seen in the English Channel, and on the southern coasts of the Moray Firth. The kind of shingle that is moved to leeward will, of course, depend on the force of the surf. Lieutenant Worthington, in his report on Dover, says that on that coast it averages 104 lbs. to each cubic foot.

Shoaling of Inclosed Harbours.—It is a mistake to suppose that, when water is inclosed by solid piers, there must necessarily be a great deposit. This misconception gives rise to the reports which are so frequently made, that basins constructed long ago have to a great extent silted up; and similar allegations are made, with as little foundation, regarding bays and natural creeks of the open sea. The only way to test the truth of such statements is to procure accurate soundings, and compare them with the original depths, when probably, in the great majority of instances, it will be found that there is no material difference from the oldest charts. These notions most likely arise from confounding *artificial* with *natural* channels. The shoaling of channels which have been dredged deeper than the original bottom forms no proper ground for predicting a similar reduction of the depths due to the natural *profile of conservancy* of the shore, which generally preserves its symmetry with remarkable persistency, even within artificial inclosing walls.

Cause of Inclosed Harbours keeping open.—Captain Calver, who is strongly opposed to close harbours, on the ground that they will fill up, makes an exception regarding small tidal harbours, which he says are kept clear by the "scavenging process" of high winds during ebbing tide, and that the "most diminutive lipper" is effective in moving the lighter kinds of deposit.

The surface-ripple described by Mr. Calver will, no doubt, have a certain effect. But there must surely be some other cause greatly more powerful and efficient, than this, to keep open our ports and harbours. The "*run*," wherever there is a *ground-swell*, and even the ordinary waves produced by a gale, are, I think, the agents which possess all the powers that are required ; for although the depth at the entrance be considerable, yet, when the wind is strong, the surface undulations become, partially at least, waves of translation. Each wave, as it enters the basin, will therefore import a certain quantity of water, which must ultimately escape seawards through the entrance, otherwise the water would stand higher within the harbour than in the sea outside. The under-surface current thus produced runs probably very near the bottom. Hence the detritus and silt that would be left in the basin, were there no such current, is carried out again into the open sea. That the quantity of water so brought in cannot be very small may be judged of from the fact that, during a gale in the Irish Sea in 1842, I counted

9·6 waves per minute, so that about 14,000 waves broke on the shore during twenty-four hours. Although each wave injected but a small portion of its contents into a harbour, it is quite conceivable that that water, returning seaward, should prove efficient as a *scouring* power, or at least in preventing the entrance of silt near the bottom. Mr. J. Wilson of Sunderland expressed the opinion some years ago in the *Engineer's Magazine* that when there is an on-shore wind there must necessarily be an off-shore current, and Mr. Cleghorn of Wick also maintains similar views. There can be no doubt, I think, of the accuracy of these opinions; but, as stated above, a *ground-swell* will generate in a close harbour an outgoing current even in the calmest weather.

Artificial Scouring.—The preservation of the depth of harbours at a level *lower* than that of the original bottom involves both uncertainty and expense. Where the deposit is confined to the space between high and low water marks, the scouring by means of salt or fresh water is comparatively easy; but where it forms a bar outside of the entrance, the possibility of maintaining permanently a greater depth becomes very doubtful. The efficacy of the scour, so long as it is not impeded by enlargements of the channel, may be kept up for great distances, but it soon comes to an end after it meets the sea. When the volume of water liberated is great compared with the *alveus* or channel through which it has to pass, the stagnant water which originally occupied the channel does not, to the same extent, destroy the momentum, as where the scouring has to be produced by a sudden *finite* impulse. In the one case the scouring power depends, *cæteris paribus*, simply on the relation subsisting between the quantity liberated in a given space of time and the sectional area of the channel through which it has to pass; while in the other it depends on the propelling head, and the direction in which the water leaves the sluice. Mr. Rendel's scheme for Birkenhead was on the former principle, which it must be recollected is only applicable where the soil is easily stirred up.

Effective Velocity of Scouring.—The quantity proposed by Mr. Rendel for scouring at Birkenhead on average spring tides was 1,600,000 cubic yards, to be liberated in the short space of three quarters of an hour.* In all cases of scouring, it is of course an essential condition that such a velocity be generated as is necessary for acting upon the soil. The largest amount of back-water will be inoperative if it has less than what may be called the *effective velocity*, ór that required for acting on the material which forms the bottom. If, for example, the discharge of the waters of a river be equalised by the construction of regulating reservoirs, there will be an actual diminution of scouring power, because a *sudden* flood could remove what the same, or even a much greater quantity of water would never effect if liberated more slowly.

The first example of artificial scouring in this country seems to be due to Smeaton, who used it effectively at Ramsgate in 1779. At Bute Docks, Cardiff, designed by the late Sir W. Cubitt, the access to the outer basin is kept open most successfully by means of artificial scouring on a large scale. The entrance was cut through mud banks for a distance of about three-fourths of a mile seaward of high-water mark. The initial discharge when the reservoir is full is stated to be 2500 tons per minute. I have known even so limited a discharge as one ton per second produce very useful effects in keeping a small tidal harbour clear of sand.

* *Port and Docks of Birkenhead*, by T. Webster, M.A., F.R.S.; London 1848, p. 27.

Duration of Scouring.—Minard holds that when a channel has to be maintained by regular and habitual scouring, the whole effect is generally produced in the course of the *first quarter of an hour.* This was made the subject of particular investigation at Dunkirk, where sections of the channel were made before and during the scour; and it was found that there was no alteration in the sectional area after the first quarter of an hour.

Reservoirs for Scouring.—Minard points out, as the best *form for scouring reservoirs,* that which will admit of the largest discharge in a given time; or, in other words, where the mean distance from the orifice is a minimum. Such a form is obviously the semicircle having the point of discharge in the centre. He also adduces the following examples of artificial scouring. At Calais the first scour removed about 100,000 cubic metres of sand = 130,800 cubic yards. At Dieppe one scour removes about 1500 cubic metres of sand = 1962 cubic yards. At Ostend, about 500 cubic metres = 654 cubic yards; and at Treport, according to Cessart, 3000 cubic metres of gravel = 3924 cubic yards, were removed. When gravel is to be displaced it is very important that the discharge should take place a little before low water, but with mud and silt, the longer the discharge continues the better, as it prevents those light matters from being brought in again.*

At Sunderland Mr. Murray designed a reservoir of 34 acres, 4½ feet deep, which is discharged in quarter of an hour by means of eight sluices of an aggregate area of 495 superficial feet, which, less 10 per cent for friction, gives 444,312 cubic feet per minute, producing a velocity of 4.0911 miles per hour, and 3.166 along the bottom. The sluices are placed at dif-

* Minard, p. 100.

ferent levels so as to act on the whole mass of water at once, and the current is visible at a distance of 2000 feet from the point of discharge.

Effective Velocities of Currents.—For the velocity required to move stones or shingle, Sir John Leslie gives the following formula :—

Where a denotes in feet the side of a cubic block of stone or diameter of a boulder, and v = the velocity of the water in miles per hour, which is capable of moving it along the bottom,

$$v = 4\sqrt{a}$$

The annexed are the results of experiments made by various observers on the size of particles which are moved by currents of different velocities :—

3 in. per sec.	= 0.170 mile per hour	will just begin to work on fine clay.
6 in. ,,	= 0.341 do.	will lift fine sand.
8 in. ,,	= 0.4545 do.	will lift sand as coarse as lint seed.
12 in. ,,	= 0.6819 do.	will sweep along fine gravel.
2 feet ,,	= 1.3638 do.	will roll along rounded pebbles 1 inch in diameter.
3 feet ,,	= 2.045 do.	will sweep along slippery angular stones of the size of an egg.
6.56 ft. ,,	= 4.472 do.	required at Havre and Fécamp to scour gravel.*

The most recent experiments are by Mr. T. Login, of which a description will be found in the Proceedings of the Roy. Soc. of Edin., vol. iii. p. 475. In these experiments, the results of which are appended, the stream seldom exceeded half an inch in depth.

* Minard, p. 106.

Nature of materials.	Rate of rushing-in water.	Velocity of Current.	
	Feet per minute.	Feet per minute.	Mile per hour.
Brick-clay when mixed with water, and allowed to settle for half an hour	.566	15	.170
Fresh-water sand	10	40	.444
Sea sand	11.707	66.22	.752
Rounded pebbles about the size of peas	60	120	1.37
Vegetable soil	—	50	.56

Brick-clay in its natural state was not moved by a current of 128 feet per minute, or 1.45 mile per hour.

Dredging.—Steam and hand dredging are in frequent use both for the formation and preservation of harbours, and the former in combination with other works has produced the most important changes in our British rivers.

At the Clyde, Tyne, Thames, Tay, Ribble, and many other rivers, enormous quantities of stuff have been excavated without interfering with the usual traffic. Mr. Ure, who designed the dredge which is probably the largest that has yet been made, and which is now in use on the Tyne Navigation, has been good enough to send me the following statement :—

"Our largest dredge here is 142 feet long, 38 feet beam, and 11 feet deep amidships, rising one foot fore and aft. The engine is 50 horse-power, the cylinder being 42 inches diameter, with the piston adapted to 3 feet 4 inches stroke, and working about 29 revolutions per minute, with a pressure of 5 lbs. steam in the boilers. The quantity of dredging that will be lifted by it this year (there is still a fortnight to run) will be about 850,000 tons, the ground being principally sand in Shields Harbour. But a good deal of interruption was experienced in falling foul of wrecks, anchors, etc. The cost of this vessel complete was about £20,000.

"The largest quantity of stuff the dredge has raised in one day, from 3 A.M to 10 P.M., was 5320 tons, which includes all detentions for shifting, etc., and stoppages. The most it has done in an hour of continuous work I have not a memorandum of, but it is about 450 to 500 tons. We send all those dredgings in proper barges to sea. It takes about eight to ten to keep her going, each carrying from 250 to 300 tons, and two steamers to tow them. Latterly I have tried a screw hopper barge, which proved successful. The distance towed will be 4 to 5 miles, and the same back for empty vessels. . . . I may mention that the cost of the plant to keep her at work—viz. the two steamers and hopper barges —will be about £25,000.

Steam Hopper Punts.—The Messrs. Symons of Renfrew have constructed large screw punts, which are now successfully used in several rivers. The screw is 8 feet in diameter, with 12 feet 6 inches pitch. The draught of water, with 300 tons on board, is 8 feet. Messrs. Henderson, Coulborn, and Co., of Renfrew, have also constructed screw hopper punts for the Suez Canal. They are capable of containing 300 tons of wet sand, and are propelled by condensing engines of 35 horse-power (nominal). When these punts were tried on the Clyde, their speed, when empty, was 10.35 miles per hour, and when loaded with 300 tons it was reduced to 9·8 miles per hour. Their dimensions are—

Length of keel and fore-rake	135 feet.
Length of hopper	50 „
Breadth on deck	19 „

The plating of vessel and hopper is $\frac{3}{8}$ inch thick, and forms three water-tight bulkheads.*

Dredging in Exposed Situations.—From special observations which I had made for me by Mr. M'Donald, resident engineer at Loch Foyle, when the dredge was at work, it appears that the dredging cannot proceed when the waves exceed or approach $2\frac{1}{2}$ feet.

* *Prac. Mech. Journal*, vol. i. 3d series, p. 73.

"When the waves rise to 2 feet or 2½ feet," says Mr. M'Donald, "we then let the vessel's head come to the winds and waves, when she rides much easier than when lying broadside. We find the punts are worse to handle than the dredge; they are so short that, with a 2½ feet wave, they would roll all the stuff off their deck in a short time; and if we took them near the dredge, they would soon either destroy themselves or the machine, or probably both. It is also severe on the dredge to work when the seas are above 2½ to 3 feet high; the weight of the buckets and ladder on the stem, and the engine and coals on the bow, tends to strain the vessel amidships. It is but seldom we have a wave even 3 feet high."

Lough Foyle is about 15 miles long and 7 miles broad, and as will be seen from the table on page 23, waves of 4 feet in height are sometimes generated during gales; yet the dredge with her punts, although unable to work, can ride in safety when exposed to waves of this height.

Mr. Ure tells me he has never tried the large Tyne dredge in rough weather, but with other dredges of the usual size he has never "done any good with more than a *two-feet sea.*"

The Removal of Silt by Pumping.—The late Mr. Duncan, engineer of the Clyde Navigation, kindly sent me the follow- extract from M. Leferme's Report of 30th September 1859, on the result of M. Gache's silt-pump of 20 horse-power at St. Nazaire:—

"The experiment has now been made on a scale vast enough to warrant giving an opinion on its merits. A longer trial might suggest the necessity of making some slight modifications in the details, but we are bound to consider the aim as having been completely attained. The silt lifted by the pumps is not, as might be supposed, mixed with water. When fairly set agoing, the density of the silt lifted with the pumps is 1.19, being evidently the muddy layer in which the sucker is at work, and which is sunk in the mud for 40 or 50 centimetres (about 18 English inches). A phial filled with the mud drawn up with the pumps, after being hermetically sealed, and allowed to stand for 36 days, had only a film of water on the top of about 0.005 in thickness.

The time required for loading the boat (which contains in its wells 225 cubic metres), transporting, and discharging, including the time lost in passing through the double lock-gates, is on the average $4\frac{1}{2}$ hours. . . . The metre cube of mud lifted from the basin and carried 1500 metres has as yet, all expenses included (excepting interest of capital), amounted to 16 centimes ($1\frac{1}{4}$d. per cubic yard)."

Mr. Duncan informed me that he found this plan did not answer at the Clyde. Although the rose was buried several feet in the mud, nothing but discoloured water was ever lifted. The material* was so porous that the water percolated freely through it, and being the lighter body was lifted by the pump, in a considerable stream.†

It is probable that in some *docks* the deposit may be of a density that would admit of its being pumped. The annexed table of deposits for different ports is from Minard:—

TABLE of Deposits in Docks. (*Minard.*) ‡

	Inches per day.
Ramsgate	.157
Hull	.118
Flessingen	.197
Havre	.276
Honfleur	.787

MEDITERRANEAN PORTS.

	Inches per annum.
Marseilles	0.236
Cassin	3.937
Ciotat	1.456
Bouc	0.394

At the old dock of Grimsby fresh water is supplied by land-streams, from which, unlike the water of the Humber,

* Its specific gravity was 1.46, that at Nazaire 1.19.

† Min. Civ. Eng.

‡ Minard, p. 95.

there is little deposit. At Cardiff, instead of using the Bristol Channel, which has much silt in mechanical suspension, the waters of the river Taff are used for supplying the Bute Docks, while the New Docks at Penarth are entirely supplied by the tide water of the Bristol Channel.

Checking Sand-Drifts.—Though the sea, with its restless waves and ever-varying tides, will always demand the greatest share of the engineer's vigilance and attention, yet it is not the only foe with which he has to contend. There are difficulties to be met on the land as well as on the sea. When high winds sweep across a large tract of barren sand, large quantities may be deposited in docks or harbours. At Ostend the sand, even when wet, has been carried by the wind to very considerable distances.

Different devices have been tried for checking sand-floods. High stone walls have never, in any instance that I have known, been found to do much good. At the harbour of Nairn a slight fence about 8 feet high, and consisting of spars of wood from 3 to 5 inches broad, and fixed from 1½ to 3 inches apart, has been found more efficient than an altogether impervious barrier. At Mullaghmore Harbour, in the county Sligo, Lord Palmerston planted a species of pine tree for checking the incursions of sand, on the advice of the late Mr. R. Stevenson, who had been struck with the vigorous growth of the *Pinus maritima major* on the shores of the Bay of Biscay. In his report to Lord Palmerston in 1839, Mr. Stevenson recommended that pine cones should be procured from France. A kind of bent grass was, on the suggestion of the late Mr. Lynch, Lord Palmerston's land-steward, planted on the side next the sea, so as to act as a protection to the pines during their first growth. The result of the experiment has been highly successful, having established the fact that the *Pinus maritima major* is, in certain circumstances, nearly as well adapted to our own coasts as to the coasts of Normandy, a fact which deserves to be more generally known.

The following information was kindly communicated by Mr. Kincaid of Dublin, as to the present state of these plantations.

"The Mullaghmore plantations extend to about 200 acres. About 80 of these were planted 25 years ago. Some of the trees are 30 feet in height, and vary from that height to about 20 or 25 feet. The remainder were planted 10 years ago, and are making fair progress. All the pine plantations from opposite Newtown Cliffony to Mullaghmore are in a most healthy condition, the trees making growths of from 12 to 20 inches each year. The storms have no bad effect on the south side of the great sandhill, but on its summit, and towards the west side, the spray and gales of the Atlantic will not allow the young trees to make any progress."

Danger of deepening the Entrance of Harbours of Small Reductive Power.—One cause of disturbance in harbours, which is often not sufficiently considered, is the inconsiderate deepening of the entrance, without making, at the same time, a proportionate enlargement of the internal area, or providing other works for counteracting the effect. As the depth of the water is increased, waves of greater height reach the entrance, and thus gain admission to the interior. At the port of Sunderland Mr. D. Stevenson recommended the removal of nearly the whole of the inner stone pier, and the substitution of works of open framework, in order to tranquillise the interior. These works, which have been quite successful, were rendered necessary by the frequent dredging of the channel at and near the entrance. Similar results have been experienced at other harbours. The Table shows some of the particulars of harbours which have suffered disturbance from deepening.

	Area in acres and decimals.	Width of entrance in feet and inches.	Rise of tide in feet and inches.	Low-water depths at entrance.		Amount of deepening in feet.
				Former depth in feet.	Present depth in feet.	
Lybster . .	2.21	80 0	10 6	2	4	2
Dunbar . .	4.01	58 0	17 0	0	2	2
Cockenzie . . (deepened from mouth inwards).	2.71	84 0	16 6	0	0	6
Sunderland . .	...	320 0	14 6	...	4	

Commercial Value of Harbours or Rivers increases as the Cubes of the Depths of Water.—It is not wonderful that the risk of admitting more *run* into a harbour should often be disregarded in acquiring a greater depth; for the commercial advantages are not proportional simply to the additional depth, but they increase in a much higher ratio. Besides there is a farther advantage due to increase of depth. For example, Mr. George Robertson has shown that by making the Albert Dock, at Leith, 2 feet deeper than the Victoria basin, there are 296 tides in the year in which there will be a depth of 23 feet, whereas at the Victoria there are only 102 tides in the year in which that depth occurs.

From an examination of the proportions of a considerable number of vessels, it turns out that, although there seems to be not much uniformity in the ratio of tonnage to draught among steam-vessels, whether propelled by paddle or screw, yet there appears on the average to be a tolerable amount of uniformity among ordinary sailing vessels constructed of timber. I have found that, although even among sailing vessels there are marked peculiarities in the build, the following simple formula, deduced from a somewhat extended examination, gives a fair general approximation to the tonnage. It cannot, however, be regarded as more than *generally* true. Where d represents the draught in feet, and t the burden in tons, and a is a constant depending on the build,

$$t = \frac{d^3}{a} \qquad d = \sqrt[3]{a \times t}$$

The ratio of draughts to tonnage has been gradually decreasing; but for the general run of timber vessels built twenty or thirty years ago, a may be taken $= 10$, and for those at present frequenting our ports, a may perhaps be assumed as $= 9$, for vessels up to 500 tons, though very many of the larger class of vessels lately built would require a factor of only about $7\frac{1}{2}$.

Draughts and Tonnage of Ships.—For vessels of heavy tonnage, the best information I have been able to obtain is Mr. James Walker's Table of Draughts, given in his report of 1835 on Leith Harbour, and stated by him as having been obtained on the most reliable authority. The Berwick Harbour Table, in Daniel's "Port Dues and Charges," published in 1842, is the most complete information I could get regarding smaller vessels.

DRAUGHTS and TONNAGES of SHIPS, from the Berwick Harbour Table 1842, and from Mr. Walker's Table of 1835.

Tonnage.	Draughts from Berwick Table.	Draughts from Mr. Walker's Table.	Draughts calculated from Formula $d = \sqrt[3]{10 \times t}$
30	7		6.7
40	7.5		7.36
50	8		7.93
60	8.5		8.4
70	9		8.87
80	9.5		9.28
90	10		9.65
100	10		10.00
110	10.5		10.3
120	11		10.6
130	11.5		10.9
140	11.5		11.18
150	12.0		11.5
160	12.5		11.7
170	13.0		11.9
180	13.5		12.16
190	14.0		12.4
200	14.5	12.5 to 13.5	12.6
300		13.0 to 16.0	14.4
400		16.0 to 17.0	15.9
500		17.0 to 18.0	17.09
600		18.0 to 19.0	18.17
700		19.0 to 21.0	19.1

The above Tables were drawn up mainly to show that we are justified in inferring from this formula that *the capacities for tonnage of different channels vary as the cubes of their depths*—a law which may be found useful when comparing the relative advantages of two navigable tracks. This result at once explains why such large sums are often cheerfully expended in securing even a single additional foot of depth in harbours or river navigations.

The following Table, calculated with the factor 10 for the smaller class of vessels, and 7.5 for the larger, gives a better idea of the class of vessels which are now generally built, though still erring on the safe side as regards the size of vessels that can pass through a channel of given depth.

Factor 10.				Factor 7.5.			
Tonnage.	Draughts.	Tonnage.	Draughts.	Tonnage.	Draughts.	Tonnage.	Draughts.
50 —	7.8	250 —	13.5	500 —	15.6	1300 —	21.4
60 —	8.4	300 —	14.4	600 —	16.5	1400 —	21.9
70 —	8.8	350 —	15.1	700 —	17.4	1500 —	22.5
80 —	9.2	400 —	15.8	800 —	18.2	1600 —	23.0
90 —	9.6	450 —	16.4	900 —	18.9	1700 —	23.4
100 —	10.0			1000 —	19.6	1800 —	23.9
150 —	11.4			1100 —	20.2	1900 —	24.3
200 —	12.5			1200 —	20.8	2000 —	24.9

Draught of Steamers in Ballast.—At the Clyde, steam-vessels of from 250 to 400 feet long draw, with their machinery on board, from 12 to 18 feet, which would give the light draught = loaded draught × .7.

Draught of Fishing-Boats.—The following Table shows the draught and principal dimensions of different classes of fishing-boats, taken chiefly from the late Admiral Washington's Report.

TABLE OF DIMENSIONS of different FISHING-BOATS.

	Extreme length.		Extreme breadth.		Draught loaded.		Do. in ballast.
	ft.	in.	ft.	in.	ft.	in.	ft.
Scotch boats, present largest size	44	0	15	6	6 to $6\frac{1}{2}$		4
St. Ives - - - -	40	9	12	3	6	6	
Isle of Man - - -	40	8	11	9	7	6	
Galway Hooker - - -	33	0	10	6	5	8	
Kinsale Hooker - - -	39	4	11	0	7	6	
Yarmouth lugger - -	52	0	14	11	7	0	
Hastings do. - -	48	0	14	11	7	3	
Deal do. - -	38	0	12	3	5	6	
Penzance - - -	40	6	12	0	6	6	
Yarmouth - - -	37	0	12	3	4	0	

Bars.—The cause of bars is not, as many writers affirm, the meeting of contrary litoral and river currents, but the heaping-up action of the waves, which thus form a kind of miniature under-water beach, whose position and height depend upon the relative forces of the waves and the outgoing river. The reader is further referred to the works of Abbot Castelli,* and to Mr. D. Stevenson's *Rivers and Canal Navigation*. The obvious cure, then, is to shelter the bar by breakwaters, which will reduce the height of the waves. In order to proportion aright the breadth of entrance to the interior breadth, the formula for the reductive power given at page 147 applies.

Vertical Scend of the Waves.—If the weather be perfectly calm, ships may enter a port, although their keels are almost scraping the bottom; but if there be any surge, the available depth becomes very considerably decreased by the vertical "scend" or plunge of the ship below the mean level of her keel.

The ruling or *available Depth* may be termed the minimum navigable depth of water reduced for "scend," which a vessel can depend upon finding at low water of ordinary springs throughout the track leading from the open sea to a safe berth in the inside of the harbour.

Mr. Meik, C.E., has obligingly communicated the method of estimating the "scend" at Sunderland. "The 'scend' of vessels," says Mr. Meik, "is the lift of the vessel, and is ascertained by taking the perpendicular space the vessel moves through, or the space between the position of the keel at its lowest to its position at its highest elevation. To arrive at what is the greatest draught at which a vessel can pass over the bar, we have generally deducted the 'scend' from the registered depth on the gauge, but properly only the space between the level of the keel of the vessel in smooth water and at its lowest level when passing through a wave should be deducted. This, I consider, would be too fine for practice. I cannot tell whether our rule is adopted at other ports; it is, however, used

* Geometrical Demonstrations of the Measure of Running Waters, by D. Beneedtto Castelli: fol. Lond. 1661.

here very generally. When passing over a wave with a 10 feet lift, some of our small colliers would scend 7½ to 8 feet, whereas a long screw collier of 180 feet in length would only scend 5 feet. We have ascertained this from actual observations. The scend is *generally* taken at *two-thirds* of the greatest lift of the wave for ordinary colliers, and *one-half* of the lift of the wave for large screw steamers—taking in both cases the lift or height of the wave to be from the lowest fall to the crest. We take principally by the eye, or in the proportions I have stated, if only the extreme height of the wave is returned. We have frequently checked this, when vessels struck slightly, and we have found it to be very correct."

Height of Summer and Winter Waves.—The accompanying diagram (Fig. 29) of the height of the waves at Lybster

REGISTER of HEIGHT of WAVES for 1852, observed at Lybster, Caithness-shire.

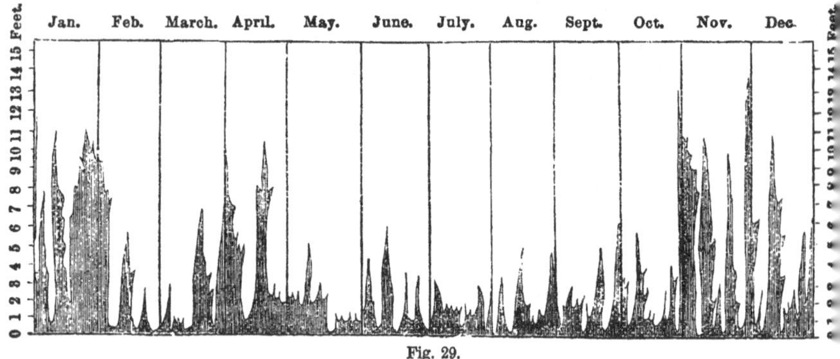

Fig. 29.

Harbour, is given for the purpose of indicating graphically the suddenness with which our eastern coast is visited by gales, and the comparative eligibility of the summer and winter months for carrying on marine works.

Lighthouses.—Every port should be provided with a lighthouse for showing the exact position of the entrance during night; and where there are outlying rocks, there should be leading lights, which, when kept in one line, conduct the mariner clear of all dangers. For a description of the different forms of apparatus, whether dioptric or catoptric, I must refer the reader to works specially devoted to the subject.* All lighthouse apparatus, in order to collect and utilise the whole of the rays, should consist either entirely of glass acting by refraction and total reflection, or else of a union of instruments of glass and metal acting by refraction and ordinary metallic reflection. Great annual loss of oil is too often entailed on harbour trustees by the use of optical apparatus, which has not been designed to meet the special wants of the locality where it is placed. In one instance, the reflectors formerly in use were replaced by a single "*holophote*" of small size, consisting of a lens with zinc reflector, which not only gives a greatly more powerful light, but has saved, according to the superintendent's returns, *two-thirds* of the oil formerly consumed. In another case where the light required to be spread over an *azimuthal* angle of 100°, one-half of the oil has been saved, and a far better and more equally distributed light produced by a small *azimuthal condensing apparatus* of glass.

Lighthouse Apparatus.—The superiority of the dioptric system is even more conspicuous at small harbours than in large sea-lights, for at the latter there is always a sufficient staff to keep the reflectors in proper polish, but this is very rarely attained in harbour lights. I have recently designed

* Rudimentary Treatise on the History, Construction, and Illumination of Lighthouses. By Alan Stevenson, LL.B., F.R.S.E. Weale, London, 1850. Lighthouse Illumination: being a Description of the Holophotal System, and other Improvements. By Thomas Stevenson, F.R.S.E. 2d edition, London, 1859. Optical Apparatus used in Lighthouses. By J. T. Chance, M.A.—*Min. of Proc. Inst. Civil Engineers*, vol. xxvi.

small dioptric holophotes and condensing lights of only one foot in diameter, which are applicable alike for harbour,

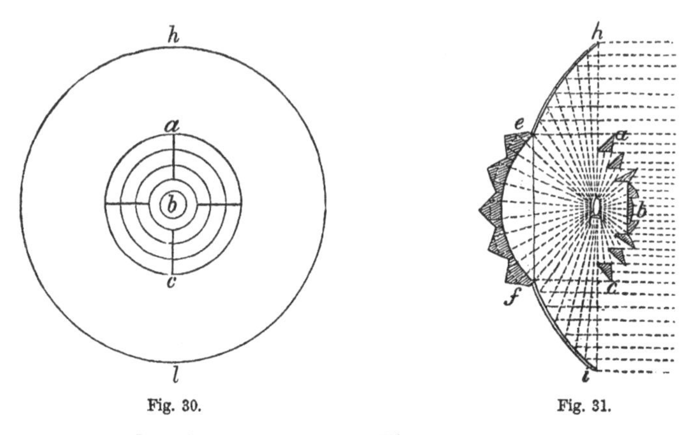

Fig. 30. Fig. 31.

steamer, and railway purposes. Figs. 30 to 33 represent the new arrangements adopted, excepting that, instead of the

Fig. 32. Fig. 33.

dioptric spherical mirror which is shown behind, mirrors of silvered copper are used from motives of economy. The rays falling on *e h f i* (Fig. 31) may also be parallelised by a combination of lenses or cylindric refractors, with totally reflecting right-angled prisms, straight when lenses are employed, and concentric with the prisms *a* and *c* when cylindric refractors are used.

Figs. 30 and 31 show in elevation and section the apparatus for operating on the whole 360°, both in altitude and azimuth, so as to produce a single beam of parallel rays; *a b c* is a dioptric half-holophote, subtending a spherical angle of 180° at the flame; *e f* is a sector of a dioptric spherical mirror; *h e f i* is a portion of a paraboloid.

Figs. 32 and 33 show in elevation and section an apparatus for parallelising in altitude, and for illuminating 180° in azimuth; *a b c* is half of Fresnel's "beehive" apparatus, *e f* the spherical mirror, and *h g i j* may be either reflecting prisms, or, as here shown, paraboloidal metallic strips, which take the form of truncated paraboloidal domes, for which can be substituted reflecting prisms and lenses, or cylindric refractors. Where the light admits of being condensed, the straight prisms of the forms suited to the given azimuthal arc are fixed outside of the other apparatus, so as to intercept the emergent rays, and thus to condense the whole 360° in altitude and azimuth into any required azimuthal arc. In 1851, in order to reduce the cost, I had made for the harbour of Morecambe, Lancashire, a holophotal apparatus, the lenticular part of which consisted of glass pressed into iron moulds, instead of regularly ground glass prisms. This economic arrangement has since been adopted by M. Degrand of Paris, who at the same time reduced to a considerable extent the thickness of the glass. I have lately had azimuthal condensing apparatus constructed on this plan for steamers' side lights, and although there is necessarily a very considerable loss of light, as compared with properly ground and polished lighthouse prisms, it is still, on economic considerations, very suitable for small harbours. In order to equalise the power of the green light on the starboard side to that of the red light on the port side of steamers, the apparatus for the green light should be made of a larger size than for the red, so as to lessen the loss of light by divergence. By adopting a larger size of apparatus, a larger burner can also be used, so as to increase the power without causing increase of loss by divergence.

I have lately designed a still better instrument than those described, called the *differential holophote,* which, *by means of single optical agents, will collect with uniform density in azimuth, the whole sphere of diverging rays into any given cylindric sector.* This is effected by using, in connection with a sector of the common fixed light apparatus of Fresnel, a mirror having in vertical section a parabolic profile, and in horizontal section an elliptic or hyperbolic profile.

Apparent Light.—There are two kinds of light to which it may be useful to refer, from their being specially applicable to harbour purposes. The borrowed or *apparent light,* as I have termed it, is a simple means of illuminating a sunk rock or other danger which is inaccessible in stormy weather. And this object is attained without requiring a lamp or any kind of flame to be placed on the rock itself. A pole or perch, carrying on its top a lantern containing different forms of optical agents (depending on the nature of the local requirements), is all that is needed on the rock itself. A beam of parallel rays, proceeding from a distant holophote placed on the shore, is thrown seawards upon the lantern on the perch, where the incident rays are re-dispersed by the apparatus, so as to produce an optical deception to the eye of the mariner, who supposes he sees a lamp burning where there is in reality none. The apparent light is also very suitable for marking the seaward ends of breakwaters or piers. In entering harbours, sometimes one pier-head must be hugged, sometimes the other, according to the direction of the wind; and so critical in stormy weather is the taking of a harbour, that even a single yard of distance may be of consequence. Those only who know, from personal experience, the anxiety that is felt on entering a narrow-mouthed harbour by night, when there is a heavy sea running, can fully appreciate the importance of descrying, at as great a distance as possible, the *exact* position of the weather pier head. But harbour lights are, from their exposed position, often inaccessible in stormy weather. At some places, too, the outer breakwaters are not connected with the shore, and can

Fig. 34.

only be reached by a boat in fine weather. The efficiency of an apparent light in such cases of difficulty has been fully tested at the entrance of Stornoway Bay, in the Island of Lewis, shown pictorially in Fig. 34, where it has been in use for the last twenty-two years, and has been favourably reported on by the captains of many vessels that have run for the anchorage at night. The beacon or perch, surmounted by the apparent light apparatus (which distributes the rays that fall upon it over an azimuthal angle of 62°), is placed on the Arnish rock, a sunk reef lying in the entrance to the bay, while the light which illuminates this beacon is placed on the land, at a distance of 530 feet. Figs. 35 and 36 show the apparatus on the perch given in section and elevation.

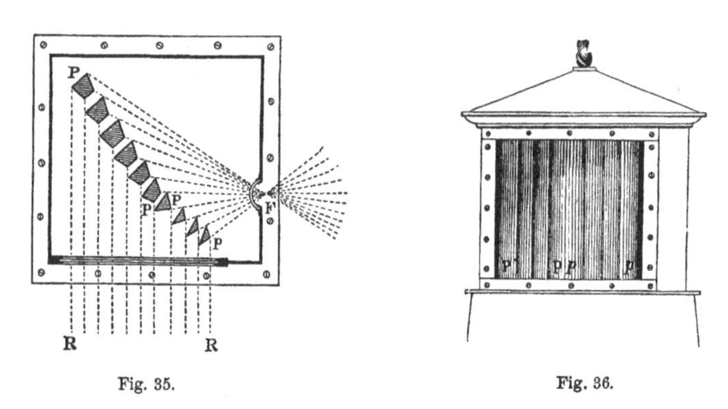

Fig. 35. Fig. 36.

Submarine Gas Pipes.—In 1851 I suggested, though disapproving of it for certain situations, that gas pipes might be laid out to any inaccessible place, the position of which required to be marked by night. Admiral Sheringham, without any knowledge of my suggestion, experimented most successfully in 1853, in illuminating a buoy by means of gas. A light on this principle has also since then been constructed on the Clyde, near Port-Glasgow, and is found to answer well. The gas, which is kept constantly burning, though with a very low flame through the day, is raised at night, and lowered in the morning by stopcocks which are placed *on the shore.*

Electric Spark.—I have lately tested a proposal which I made in 1851, for illuminating beacons or pier-heads by means of electricity conveyed through submarine or suspended wires connected with the shore. The induction spark is produced in the focus of optical apparatus.

Suspension Piers.—The suspension principle which has been found so convenient and so economical in spanning valleys, where the undertaking would otherwise have proved impracticable, has also been occasionally resorted to in marine architecture. Where the beach is long and shallow, a harbour or pier-head of timber or masonry, erected at or near the low-water mark, can be easily and cheaply connected with the shore by means of a suspension bridge. The inducements to adopt the suspension principle are, its economy and the free passage it affords to the currents, which in this way are prevented from forming accumulations of sand, silt, or gravel. These advantages are, however, reduced by the perishable nature of the structure. The late Sir Samuel Brown erected two chain piers, the one at Brighton, and the other at Newhaven, near Edinburgh, both of which are still in existence.

Timber Ponds and Discharge of Timber.—Timber ponds hold about 1000 loads per acre. For example, at Grangemouth, with an area of 12 acres, 12,000 loads of timber can be stored. At Belfast, the ponds are from 150 to 230 feet wide at water surface, which, with the logs lying at the sides, allows a passage up the middle. The rafts are 40 feet wide, eighteen inches being allowed to each log.

A vessel will discharge in the usual way from 150 to 260 logs in a day, and on the Clyde the cranes for discharging timber put out about 500 tons of timber in a day.

Advantages of two Entrances to a Harbour.—In every situation where it is easily practicable to make two entrances to a harbour, it will be found well worth the additional expense, provided they can be so placed, that the one shall be available when the other has become difficult of access. In harbours which have but one mouth, vessels are often detained for a great length of time when the wind blows in such a direction as to throw a heavy sea into the entrance; whereas if there are two mouths situated as we have supposed, vessels are at once enabled to take their departure from the sheltered side. At the port of Peterhead, the north and south harbours were some years ago united by a canal, and there the advantage has been of the most marked description. Vessels can now clear out as soon as loaded, either by the north or south mouth, according to the state of the sea. Some caution is necessary, however, in forming these communications, as the *run* is apt to extend from the one basin to the other unless there be a considerable area; and where the tides are strong the currents may also prove troublesome.

Mr. Mitchell's Screw Piles.—The ingenious invention of Mr. Mitchell of Belfast, by which piles can be screwed into the ground, has been applied by him to harbour purposes. In quicksands and silts it had been often found impossible to drive piles satisfactorily, and after they had been driven they have been known to start up again. Whereas piles fitted with screws on their ends are not only easily put down, but take a remarkably firm hold of the ground. Screws have been used as large as 4 feet in diameter, and have been made to penetrate clay and sand to the depth of 26 feet. The pier at Courtown, in the county of Wexford, is constructed on this principle, which has also been successfully employed for lighthouses and beacons.

Screw Moorings.—One of the most valuable of the applications of the screw is that for the mooring of vessels in roadsteads, rivers, and harbours. Screw moorings vary in depth from 8 to about 18 feet, depending on the tenacity of the material and on the strain to which they are to be subjected.

Methods of economising time in the despatch of Vessels.—Where commerce is prosecuted with the energy which is now so common in our own and other countries, the value of time becomes greatly enhanced, and whatever tends to shorten the length of voyages, and the time of loading and discharging vessels, is justly looked on as a public benefit. The labours of Maury, for example, who has so skilfully studied the general laws which regulate the direction of the winds and the currents of the ocean, and which have resulted in much economy of time, have been universally and gratefully acknowledged by the shipping interests. Since the publication of his wind and current charts, the voyage from Washington to the Equator has been abridged by 10 days. That from California, which used to occupy 183 days, does not now extend beyond 135. The average time of a ship between England and Australia, which used to be 124 days for going and as many for returning, is now reduced to 97 for going and to 63 for returning. The actual saving in money which has resulted from Maury's charts was estimated in 1834 at 2,250,000 dollars for the shipping of the United States alone.*

When so much has been done to abridge the length of voyages, it becomes an object not less worthy of attainment that proper despatch should be secured in the discharge and loading of cargoes after the vessel comes into port. "Had we been out a week or ten days earlier," says a witness on the Birkenhead Dock Bill,† "we should have got a freight which other vessels obtained. And at particular seasons of the year it is still more important, because, when bound on long voyages, the loss of a week is of very great importance. At a critical season it involves the necessity of the vessel taking a longer route, a much longer distance and longer in time, perhaps a month, that would very often throw her into a contrary monsoon in returning."

Sir W. Armstrong's Hydraulic Apparatus.—The principle

* Les Phenomenes de la Mer, par Elie Margollè : Paris, p. 70.

† Port and Docks of Birkenhead : Lond. 1848, p. 35.

of applying hydraulic pressure for opening the gates, bridges, and sluices, the capstans for hauling vessels out and into dock, the discharging of ballast, the loading of coals, and the shipment and discharging of general cargoes, has now been successfully adopted at many harbours. Its use is, however, only warrantable where there is a great amount of traffic, and especially at spring tides when the sea remains only a short time at the same level. At the Victoria Docks, where it is employed, Mr. Bidder * mentions that, on the 12th May 1858, 41 craft and 17 ships, or 11,711 tons, came in at one tide. In the month of April 1859 the number of

Craft entering the harbours . . .	1229
" leaving	1288
Ships entering	250
" leaving	258

or an aggregate of 2517 craft and 508 ships during the month. The gates, which are 80 feet span, are opened in less than 1½ minute. At Sunderland the *accumulators* are equivalent to a head of 600 feet, and the engine is of 30 horse-power. The gates (60 feet) are opened in about two minutes, and closed in about the same time, by one man at each gate. A wrought-iron bridge 16 feet wide, and including counter-weight equal to nearly 200 tons, is raised vertically 18 inches and drawn back in about 2½ minutes.† At Swansea the accumulators are equivalent to an effective pressure of 750 lbs. per square inch, and there are three high-pressure engines of 80, 30, and 12 horse-power respectively. The time employed in either opening or closing the gates is about *two minutes and a half*, which is the shortest period consistent with safety. The wrought-iron swing-bridge can be opened or shut in *one minute and a half*. The ballast cranes, which are dis-

* Min. Civ. Eng., vol. xviii. p. 486. † Ibid. vol. xv. p. 442.

tributed round the dock, can each discharge from 350 to 400 tons in the day. The quantity of coal that can be shipped is about 1000 tons per day, and the effective quantity of water required for the port is 21,050 cubic feet per week.*

The saving of time effected by this method is very decided; for at Liverpool, according to Mr. A. Giles,† gates of 70 feet require 20 minutes and 6 men on each side to open them. At Peterhead there is a swing-bridge with double roadway, from designs of Messrs. Stevenson, which, though not on the hydraulic principle, is moved with great ease. Each leaf weighs 91½ tons including 13 tons of ballast, yet it can be opened with one hand. The time required to raise the strut frame and open the bridge, with one man on each side, is about *two minutes and a half.*‡ The large new bridge at Leith, designed by Messrs. Rendal and Robertson, and worked by the Armstrong apparatus, is 120 feet span, length of girder 214 feet, weight moved 750 tons, and is opened in 1½ minute.

Fenders.—For the mutual protection of ships and the quay-walls at which they lie, fenders of timber are fixed to the masonry at such distances apart as to suit the length of the vessels or boats that frequent the port. They prevent the grinding action of the vessels' sides against the masonry, while they distribute the pressure over a large surface, and thus prevent undue force from coming on any single stone. Fig. 37 (in side elevation) shows a common method of attaching the fenders to the stones at top and bottom. Figs. 38 and 39 represent in plan and front elevation another method which I have employed for many years. In each of the iron palms which embrace the pile a square hole or slot is made, of

* Abernethy, Min. Inst. Civ. Eng., vol. xxi.

† Min. Civ. Eng., vol. xviii. p. 482.

‡ J. Laurenson Kerr, Trans. Roy. Scot. Soc. of Arts, vol. iv.

a size considerably larger than the wedge-shaped cutter which secures the fender. A hole is also bored through the fender of a larger size than is needed for the cutter. When the fender is to be fixed it is placed upright against the wall and between the palms at top and bottom. The cutter is then driven through the palms and the fender, until it wedges the fender close to the masonry. As the outer edge of the cutter bears hard against the outer edge of the slot in the palms, while the inner edge bears upon the inner side of the hole in

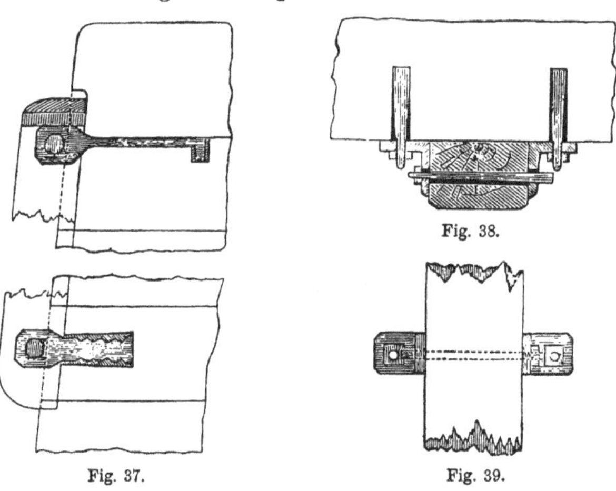

Fig. 38.

Fig. 37. Fig. 39.

the fender, a close contact of the fender with the masonry is thus effectually produced, and their contact is preserved by a nut which is screwed upon the smaller end of the cutter. After the timber has decayed or has worn out, the fender is easily taken out by unscrewing the nut, when a new fender can be put in its place. The advantage of this plan is the certainty which it provides for the uniform bearing of the fender on the quay-wall, whereas in the common arrangement it is difficult to effect this, and at the same time to get the bolts to fit closely into the holes. Where both these objects are not attained by accurate fitting, the whole force must come against the iron palms, which are from this cause frequently broken.

Loading of Coal and Coal Staiths.—It is of importance in discharging coal into ships that the fall should be as small as possible. At Bramley Moor Dock, Liverpool, the Wigan Coal is discharged by railway 18 feet above the quay. At Sunderland Mr. Meik finds that 20 feet above the quays, as at the South Dock, is too small for the shoots, and prefers 36 feet. At Penarth the height is 22, and they are able to lift 180 tons per hour. At Greenock, where steam cranes are used, they load about 500 tons a day. At Middlesboro' they can discharge 150 tons per hour, but 106 tons are as much as can be trimmed on board. At Cardiff, where there is no hydraulic machinery, they load 100 tons an hour. The distance between the staiths is 180 feet.

At the Tyne Docks Mr. Harrison states that on one occasion, when a vessel had very long hatchways admitting of several waggons being simultaneously discharged, 420 tons of coal have been shipped in 55 minutes, a feat which shows the admirable mechanism employed.

Ballast Cranes.—At Swansea, with hydraulic machinery, they can discharge from 350 to 400 tons a day, and those at Penarth load at the rate of 60 tons an hour.

Tonnage of the greatest British Ports.—The following table shows the total amount of imports and exports of the greatest ports of the United Kingdom. It is from the Nautical Magazine of 1873, in a paper entitled, "Our Great Ports," and is stated to have been obtained from trustworthy sources.

	Tons.
1. London	11,595,482
2. Liverpool	11,321,145
3. Tyne ports	7,425,698

		Tons.
4.	Cardiff	3,174,905
5.	Sunderland	3,139,699
6.	Glasgow	2,641,477
7.	Hull	2,634,018
8.	Dublin	2,591,690
9.	Belfast	1,984,080
10.	Swansea	1,710,831
11.	Southampton	1,644,239
12.	Bristol	1,564,697
13.	Hartlepool	1,578,341
14.	Newport	1,367,969
15.	Leith	1,134,123
16.	Cork	1,007,437

Mooring-Pawls.—Stone, cast-iron, and timber, are used for mooring-pawls. Fig. 40 represents a cast-iron pawl, and Fig.

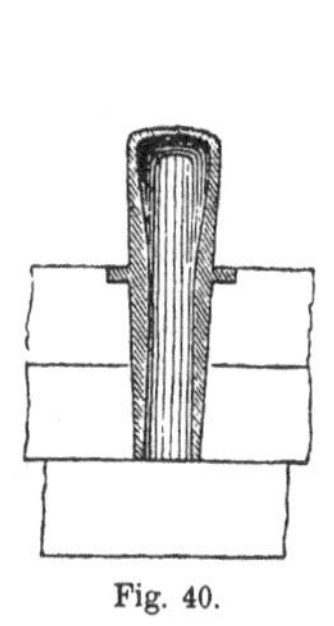
Fig. 40.

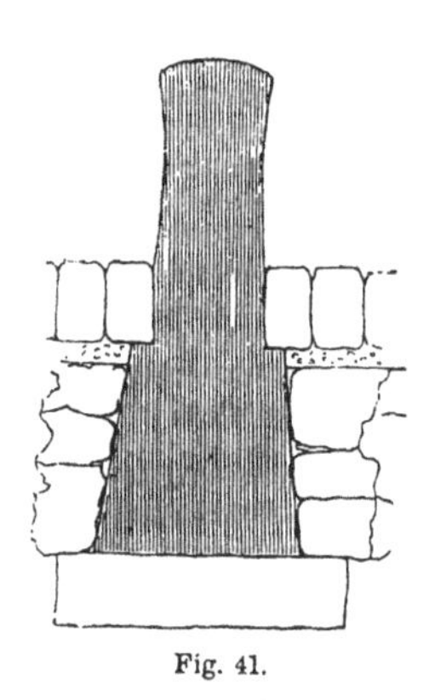
Fig. 41.

41 is one formed of stone. The best materials for the latter are granite, limestone, or other tough material. Cast-iron pawls have been perforated on the top, so as to give an exit for compressed air at piers which are exposed to a heavy sea.

Landing Platforms or Floating Stages.—A landing platform or floating stage was first used, so far as I am aware, in the Mersey, by Sir W. Cubitt. It is 507 feet long and 80 feet wide, and weighs about 2000 tons. The platform, which is of timber, rests on 30 malleable iron pontoons, each 80 feet long, 10 feet wide, and 6 feet deep, drawing when loaded 3½ feet. Each pontoon can be detached and removed for painting or repair. The landing stage is connected with the shore by two hollow girder bridges 154 feet long, and having joints at each end, admitting of horizontal and vertical play, so that the stage can rise and fall freely with the tide, without disturbing the traffic. The moorings run from the quay under the bridges to the deck, thereby rendering unnecessary the use of anchors or chains, which might prevent vessels from coming freely alongside.

Capstans.—Capstans worked by handspikes are a common and useful adjunct at the entrance to a harbour. At the one which is placed in the port of Honfleur near Havre, I noticed a simple expedient, which has been adopted for preventing ships' cables from *riding* when being *coiled* on the drum. It consists of an iron ring about four inches in width and about two inches thick, which encircles the capstan, and which is gradually pushed upwards as more and more of the cable is coiled on the drum or barrel. The heavy ring is thus constantly pressing upon the cable, so as to prevent it from rising too high on the barrel.

Caissons, employed instead of gates, are the invention of the late General Bentham, and are now frequently used both for wet and graving docks. These caissons, constructed of malleable iron plates, are sometimes made in the form of a ship's hull, and sometimes have vertical sides. The caisson fits into checks made for its reception in the side walls of the entrance to the dock. They require to be very carefully ballasted, and to be most accurately fitted to the masonry on the sides and bottom. At high water they are floated into the grooves, and are then scuttled by admitting the water into the interior chamber. When a vessel has to leave the dock, the water is again pumped out, when the caisson is floated out of its place and taken into the river or harbour, so as to be out of the way of the vessel which is leaving. General Bentham says they are cheaper than gates, occupy less room, are more easily repaired, and the same caisson may be used for different places at different times, while they answer for roadways and require less labour for opening.

The caisson at Keysham was designed by Mr. Scamp, Deputy-Director of the Admiralty works, and is 80 feet wide and 43 feet deep, with an air-chamber at the bottom. When raised a few inches above the bottom, the caisson is drawn back into a recess or chamber in the side walls. The total weight is 290 tons, and the deflection was $\frac{7}{8}$ inch near the bottom when the pressure over the whole surface was 2000 tons. The time for opening and closing the entrance at Keysham is 10 minutes and 8 minutes respectively. The cost was about £10,000.*

The wrought-iron caisson and folding bridge at the Garvel Graving Dock, Greenock, was designed by Mr. Kinipple, and has been patented by him. He has given me the following description:—"The bridge is of sufficient strength to admit of a locomotive passing over it. Instead of being floated out of its berth, at great loss of time and labour, as has hitherto been usual with ordinary caissons, upon vessels entering or leaving a dock, Mr. Kinipple's caisson is carried upon trollies running upon plate rails, or upon rollers fixed on the floor of the caisson chamber; and, by means of a small hydraulic apparatus, is drawn into or out of a chamber or recess under

* Min. Civ. Eng. vol. xiii. p. 444.

the quay as may be required. All the invert and stop quoin faces of the entrance against which the caisson with its teak meeting faces, slides and abuts, are of polished granite. The bridge, by coming in contact with curved plates, lowers or raises itself during the process of opening or closing. Among the many advantages to be obtained in connection with this invention, it may be mentioned that the caisson may be opened or closed in a few minutes, at any time of tide, and in almost any sea or weather, or during a considerable current through the entrance; it also avoids the cost of heavy swing bridges, with expensive foundations, opening and closing machinery, etc. The caisson may be also floated in the ordinary way, and removed to any graving dock for repairs, or be placed outside the entrance to act as a coffer-dam in the event of its being necessary to get at the cill of the dock or other parts of the entrance works below low water."

Steam Cranes.—At Glasgow, the 60-ton crane has a sweep of 42½ feet; the 40-ton crane has a sweep of 34½ feet; and the 30-ton, 23½ feet. The largest cranes require, according to Mr. Deas, a sweep of 47½ feet, and should be placed 16 feet from front of the quay.

Seventy-ton Crane at Greenock.—One of the largest steam cranes that have as yet been made was erected under the superintendence of Mr. W. R. Kinipple at Greenock, who has contributed the following information regarding it:—

"The crane is one of the first completed of a number of large cranes which Messrs. Taylor, Birkenhead, have in hand for various Harbour Trusts in the country—two being in course of erection at Plantation and Stobcross, Glasgow. The crane has a radius of 57 feet, and its total height is 71 feet. It was tested first with a load of 70 tons of pig-iron, which, with a pressure of 40 lbs. of steam, was lifted at the rate of two and a half feet per minute, being a minute less than the guaranteed time. It made a complete revolution in five minutes, or one minute less than the guaranteed time. Afterwards the crane lifted a load of 91 tons, being one ton more than the guaranteed test load, and raised it about 12 inches. The lifting chains are 1¾ inch diameter, and have been tested to 78 tons, which, with four plies, would be equal to 312 tons. Both cheeks and jibs are of wrought iron. Besides the principal crane, an auxiliary is provided for 10 tons and under. The auxiliary power lifts 10 tons at the speed of 10 feet in height per minute.

"The roller-path and the 60 rollers are made of a material consisting of cast-iron and steel, with a small proportion of hæmatite iron, being, it is believed, about 16 per cent stronger and considerably more durable than the best ordinary cast-iron. No cast-iron is subject either to tension or torsion in any part of the structure; the main frame, jib, etc., being, as already stated, of wrought iron. There are two steam engines, one for hoisting and one for turning, so that both operations can go on simultaneously. The hoisting engine has cylinders of 10 inches and 16 inches stroke, and those of the revolving engine are 8 inches and 12 inches stroke. They are supplied from one vertical boiler. The most important strains are as under:—

	Tons.
"Tension on centre pin, with test load, and allowing for effect of overhanging structural weights on jib, block, etc.	349
Tension on each holding-down bolt, do.	58 16
Compression on jib, do.	327
Tension on tension rods	259
Compression on front roller-path, supported by 10 rollers, allowing in like manner for effect of overhanging weights, but excluding the weight of the body of the crane	460"

Iron Piers.—Piers of cast and of malleable iron are now frequently employed. Examples of these may be seen at Scarborough, Southport, Portobello, and at many other parts of the coast, and have been found to answer even where there is a considerable sea. At Scarborough pier I measured waves about 6 feet high, which struck upon the end of the pier.

The pier at Southport, described* by Mr. H. Hooper, is of cast iron, and the mode of sinking the piles was peculiar. "The piles proper or lowest lengths of the columns are cast in lengths of 8 feet and 10 feet, and are sunk into the sand to the depth of 7 feet and 9 feet respectively. They were provided with circular discs 1 foot 6 inches in diameter, to form a bearing surface, and a small hole being left in the centre, a wrought-iron tube, 2 inches in diameter, was passed down the inside of the pile and forced about four inches into the sand, a connection being made by means of a flexible hose between the top of the tube and a temporary pipe connected with the Water Company's mains, and extended as the sinking of the piles proceeded. A pressure of water of about 50 lbs. per inch was thus obtained, and this was found to be sufficient to force the sand from under the disc. Each disc was provided on the lower side with cutters, which, on an alternating motion being given to the pile, loosened the sand. The piles were gradually lowered, and guided by a small ordinary piling engine. When the pressure of water had been removed about 5 minutes, the piles settled down to so firm a bearing, that when tested with a load of 12 tons each no signs of settlement could be perceived."† The cast-iron columns are 7 inches in external and $5\frac{3}{4}$ inches in internal diameter. All the piles, to the number of 237, were sunk in 6 weeks, being at the rate of between 6 and 7 in 24 hours.

* Min. Inst. Civ. Eng. vol. xx.

† *Ibid.* p. 293.

The Clevedon pier is of malleable iron. J. W. Grover* states that each upright consists of two Barlow rails weighing 80lbs. to the yard, rivetted back to back, and having a total section for each 100 feet span of 64 inches. They are braced together by diagonal tie-rods from $1\frac{3}{4}$ to $2\frac{1}{8}$ inch in diameter. The lower portions of the piles below low water are of solid wrought iron, 5 inches in diameter, shod with cast-iron screws 2 feet in diameter, and were screwed down till a $4\frac{1}{2}$-inch rope passed round 6 feet capstan bars parted with the strain. They penetrated the ground to depths varying from 7 to 17 feet, and though made with a thread of 5 inches in pitch, seldom descended more than $2\frac{1}{2}$ inches or 3 inches in a turn. The solid pile-stems are connected with the Barlow rail piles by cast-iron shoes. Where rock occurred holes were jumped, and a 4-inch wrought iron bar was inserted and secured by a jagged key. A shoe to receive the Barlow rail was fitted and keyed on this, and the remaining space was caulked with iron cement. The length of the longest pile is 76 feet. The level of roadway is 16 feet above extreme high water, and the height above the ground at the pier-head is about 68 feet.

Advantages of Government supplementing Local Funds.—In concluding these remarks on Harbours, it may not be out of place to state that the want of sufficient funds occasions a great national loss in the construction of many of our ports. The history of a large number of works which have been erected by private or local enterprise presents but a record of the building of piers at one period when funds were small; and of taking them down again at another, when the trade had increased, and more room and accommodation were required. The difficulty of procuring capital for schemes, how-

* Min. Civ. Eng. vol. xxxii.

ever beneficial in their tendency, and however likely to be ultimately productive, is fully established by the early history of many of our now flourishing ports.

Mr. Webster justly affirms on this subject "that if it be a true principle of commercial policy that docks should be in advance of the actual wants of trade, that stations should be freely afforded on the highway of the seas in which the products, artificial and natural, of all nations may be collected for interchange and distribution, it follows that capital must be employed on an object which, for the time being, or on a limited view of the case, may be regarded as unproductive; and hence arises the difficulty, in questions of this nature, of providing ample dock accommodation for a rapidly-increasing and variable state of commerce at the minimum rate of charge consistent with profitable investment. The history of the dock estate of Liverpool, and of the struggles and questions to which its constitution has necessarily given rise, will afford curious illustration of the difficulties with which this question is surrounded." *

The want of sufficient funds often prevents the original works from being carried within deep water, and in consequence the most expensive part of the protecting breakwater is put down just in the very place which has afterwards to be converted, at great expense, into a deep-water access or berthage. Sometimes, indeed, a whole line of pier is, from motives of economy, placed in such a manner as to interfere most materially with what might have been by far the best and safest berths for shipping, so that in the future extension of the works part of the old harbour has to be demolished. Want of a proper marine survey has also led to very serious errors in the

* The Port and Docks of Birkenhead, by T. Webster, M.A., F.R.S. London, 1848.

position of piers. It is most important, therefore, that in all designs for harbours the principle of making improvements and extensions *anticipative* should be clearly kept in view.

To such an extent has this system of partial and limited improvements prevailed, that were an engineer called on to value many of our works as they exist at present, his estimate, however fairly and fully made out, would fall far short of the actual cost. For it would proceed on a measurement of what he sees, while the actual cost would include the building of piers and jetties which had ceased to exist. For these reasons we conceive there could hardly be a more advisable expenditure of the public money than a system of grants, on a liberal scale, for supplementing the local funds. With such aid the authorities on the spot would be enabled to protect and improve the existing physical advantages which the shores possess, by preventing the construction of proposed improvements on too narrow a scale. But a comparatively slight increase of the means would, in many instances, inclose a great additional area, and secure a deeper access with superior internal tranquillity ; the want of which cripples the trade, and becomes a subject of lasting regret to all frequenting the port.

It is gratifying to add that since the above remarks regarding Government aid appeared in the article "Harbours" in the *Encyclopædia Britannica*, a bill, by the Right Honourable T. Milner Gibson, then President of the Board of Trade, was brought into Parliament and passed, for granting for the construction of harbours, loans under £100,000, at the rate of 3¼ per cent. This Act has already, at many places, conferred most material advantages, in which the nation at large will doubtless eventually participate.

REFERENCE TO PLATES.

Plates I. II. and III. represent different modes of finishing pier-heads and of terminating talus walls at various harbours. The arrows denote the directions in which the heaviest waves strike the piers.

Plates IV. V. and VI. show cross sections of different tidal piers of masonry ; while Plates VII. and VIII. show similar cross sections of breakwaters in deep water.

Plates IX. and X. are cross sections and elevations of tidal piers and quay walls of timber.

Plates XI. and XII. refer to the breakwater at Wick, described at page 45 ; the former showing the block of 1350 tons which was dislodged by the waves, the latter representing a photograph, kindly furnished by Mr. Johnston of Wick, of the waves when striking upon the breakwater, the parapet of which is 21 feet above the sea.

Plate XIII. shows all the important sea lighthouses hitherto constructed in exposed situations, drawn to the same scale, and referred to at page 105. The hard line drawn across the Plate represents the level of high water of a spring-tide, and the dotted line shows the level above high water at which the masonry was dislodged at the Dhuheartach Lighthouse, Argyllshire, and which is the same as that of the light-room of Winstanley's lighthouse on the Eddystone.

Plate XIV. is the gates of the Londonderry Graving Dock, as designed by Messrs. Stevenson. Fig. 1 shows elevation ; Fig. 2 cross section ; Fig. 3 shows the mode in which the beams were built together. The other diagrams show details of friction-roller, etc., as designed by Messrs. Rendel and Robertson.

Plate XV. shows the cylindrical Victoria Dock gates, designed by Mr. G. P. Bidder, and as given in the Proceedings of the Institution of Civil Engineers.

Plate XVI. Fig. 1 to Fig. 9 show details of the dock gates designed by the late Mr. J. M. Rendel, for Great Grimsby. The drawings are from the Minutes of the Institution of Civil Engineers. Figs. 10 and 11 are the anchor plates for Londonderry gates, and Figs. 12 and 13 those of the gates at Great Grimsby.

Plate XVII. shows quay walls at Glasgow Dock and Morecambe Harbour, and at Plantation Quay, Glasgow, in plan and section, referred to at page 203.

Plate XVIII. gives the section of the new breakwater at Aberdeen, by Mr. W. Dyce Cay, and of the Delta of the Danube, by Sir Charles Hartley. In the section of Aberdeen Breakwater A is the foundation of concrete deposited liquid in bags, in masses up to 18 tons. B are concrete blocks, from 12 to 25 tons, built regularly in courses, with good bond from the foundation-concrete up to above low water. C is concrete deposited liquid *in situ* in frames, in masses of 600 to 1300 tons, extending from above low water to 11 feet above high water ordinary spring-tides. D are 100-ton blocks deposited liquid in bags, each bag holding 100 tons, forming the apron of the work. The depth of water is 20 feet at low water of spring-tides, and 32 feet at high water.

Plate XIX. shows a cross section and plan of Anstruther Harbour.

Plate XX. represents, in plan and cross section, the Graving Dock at Londonderry, erected in 1862, showing the system of underground drainage for keeping the floor dry.

INDEX.

Printed by R. & R. Clark, *Edinburgh.*

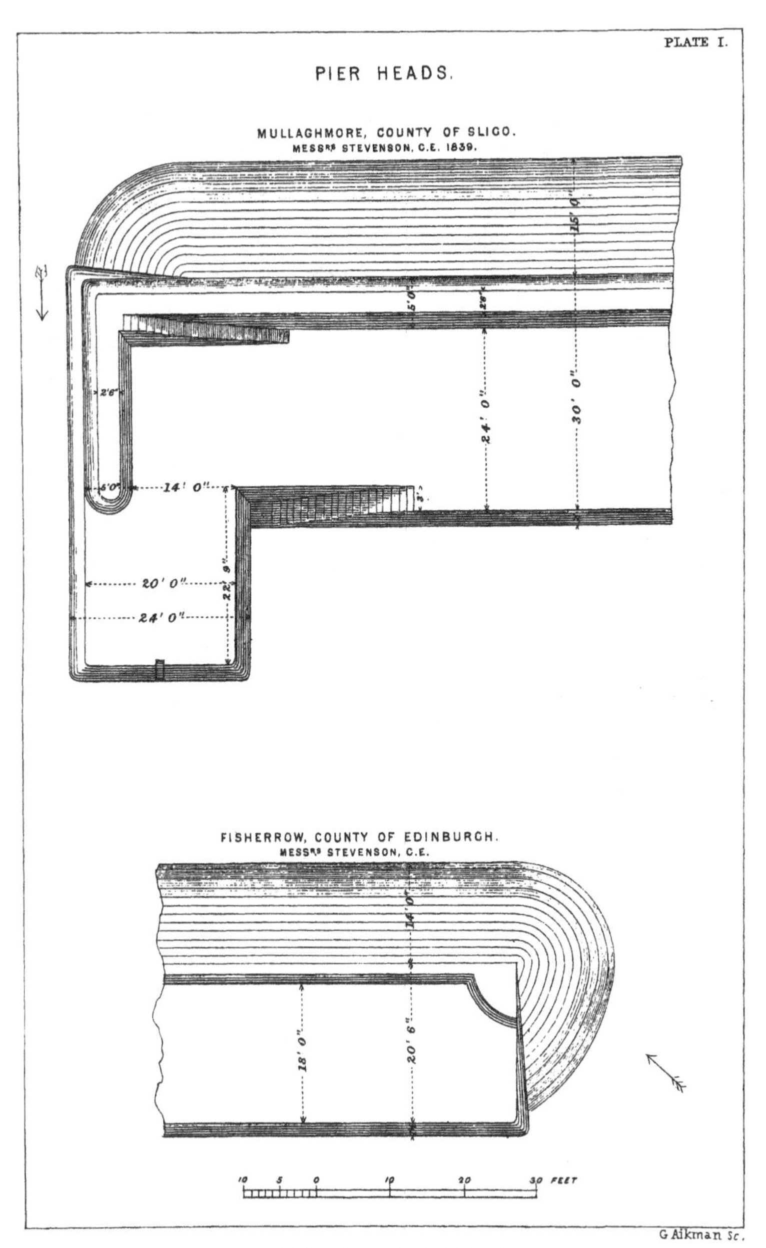

Published by Adam & Charles Black 1874.

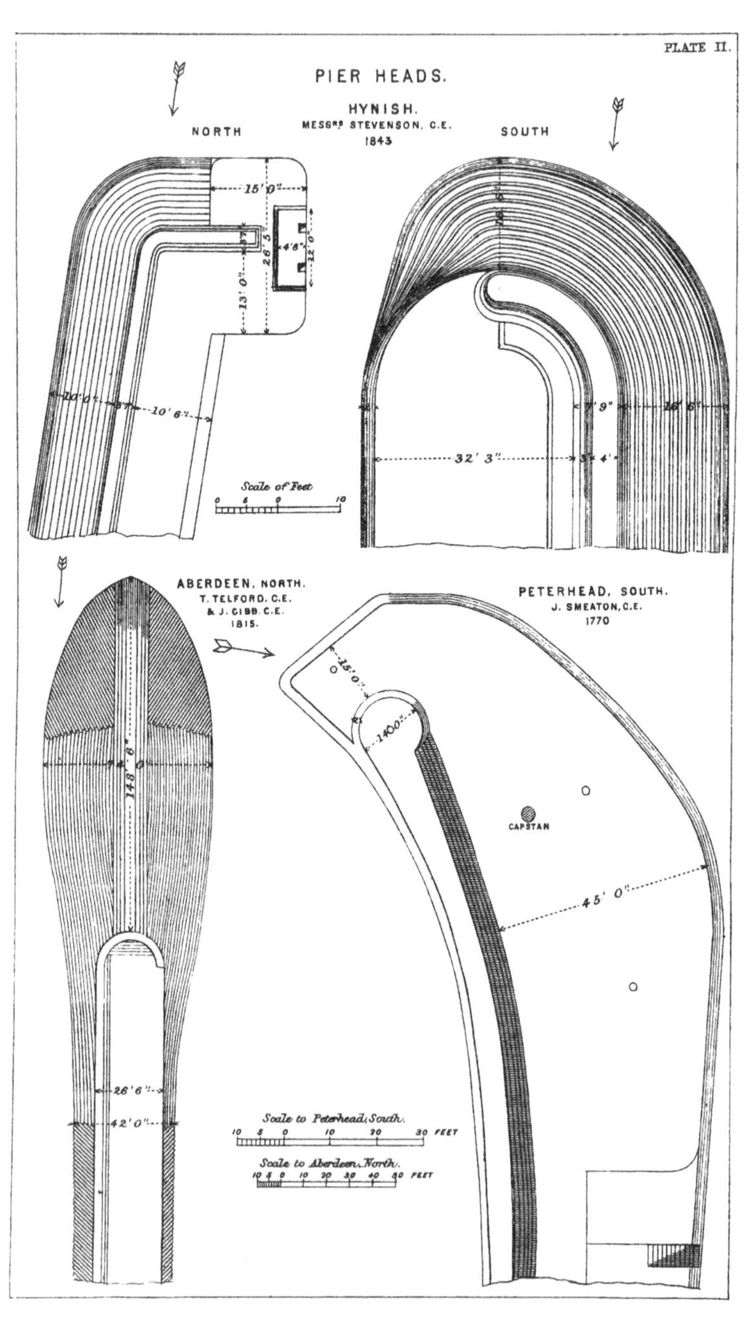

Published by Adam & Charles Black 1874.

PLATE III.

PORTPATRICK, COUNTY OF WIGTON.
J. RENNIE C.E.

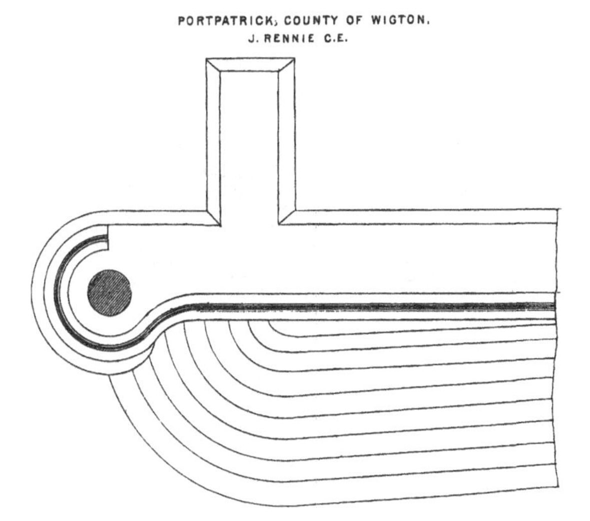

SARCLET. Co. CAITHNESS.
J. MITCHELL C.E.
1833.

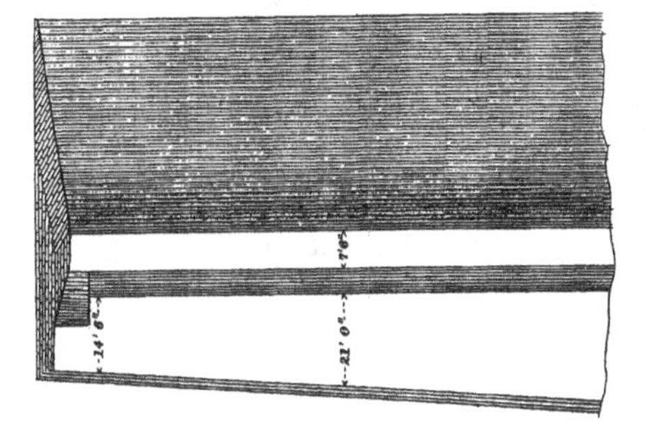

Published by Adam & Charles Black, 1874.

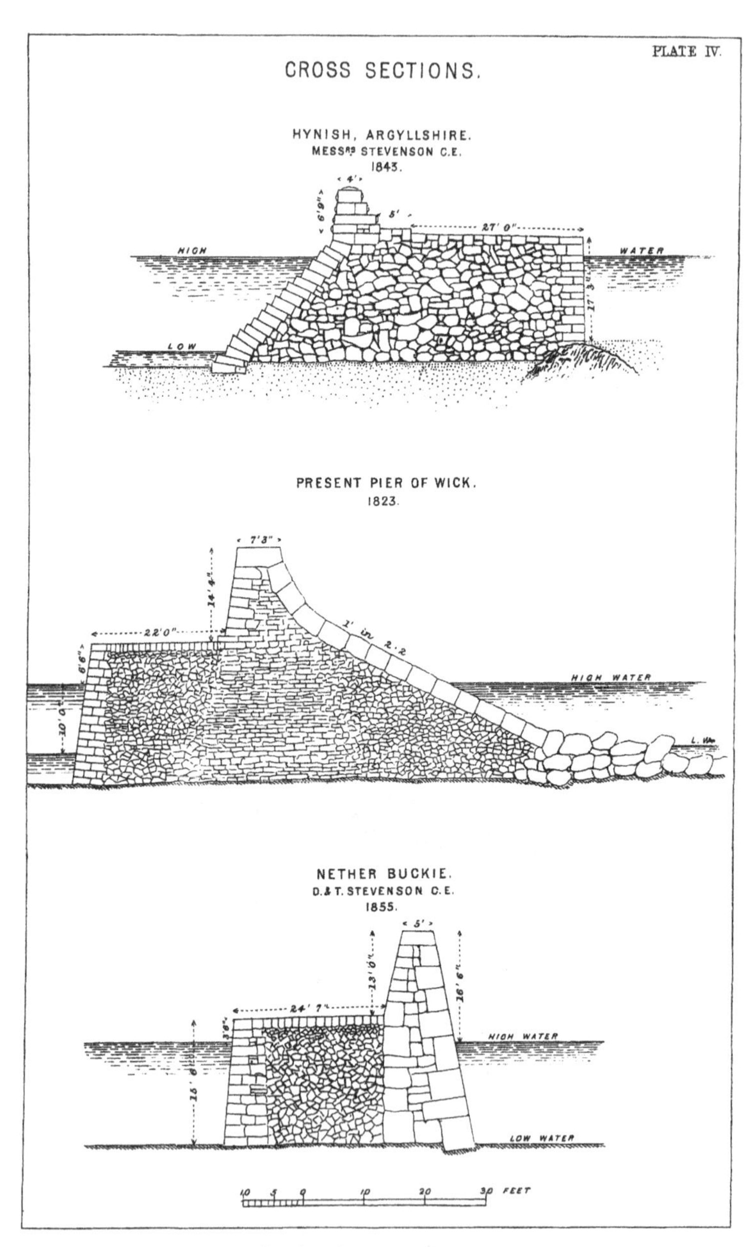

Published by Adam & Charles Black, 1874.

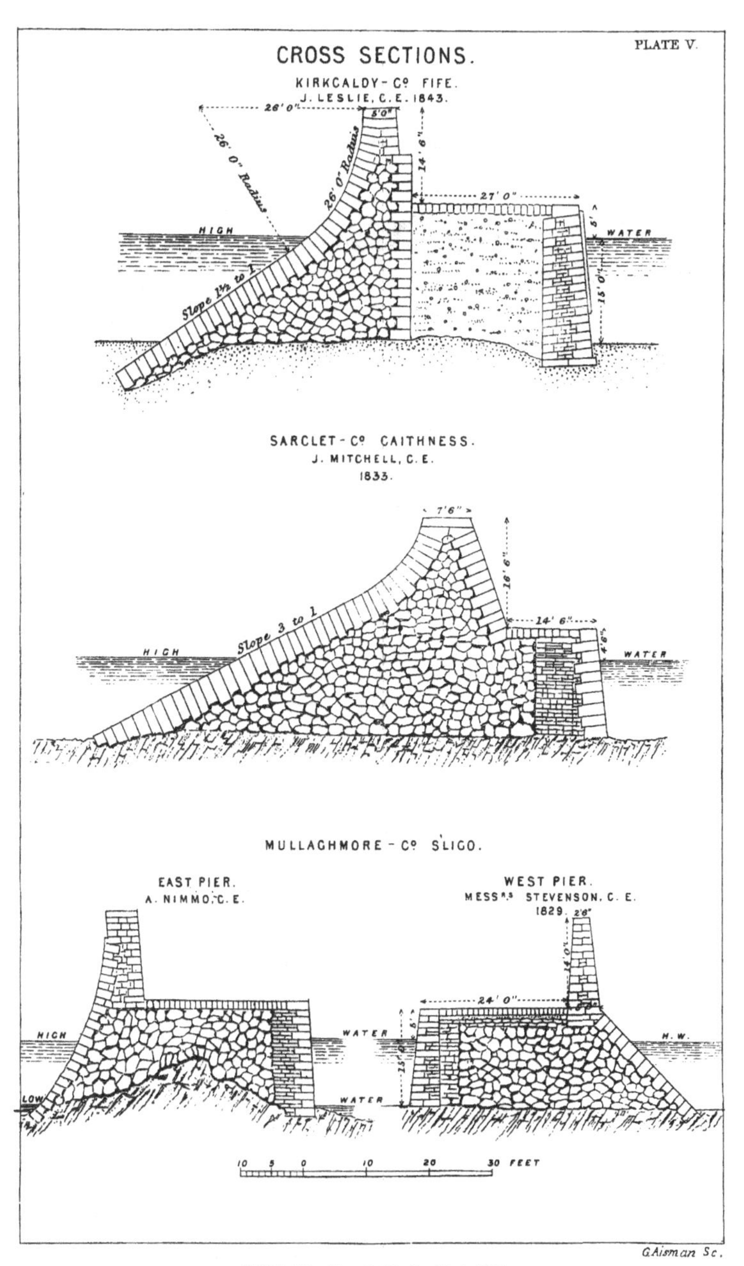

Published by Adam & Charles Black, 1874.

NORTH PIER, SUNDERLAND.

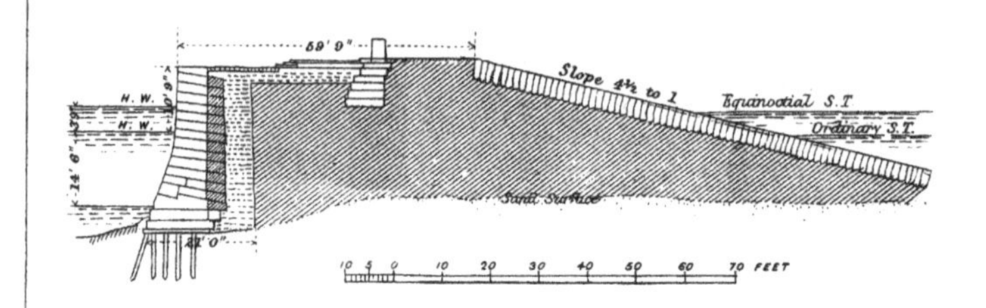

NORTH EAST PIER, SEAHAM,
EASTERN EXTENSION,
W. CHAPMAN. C. E.

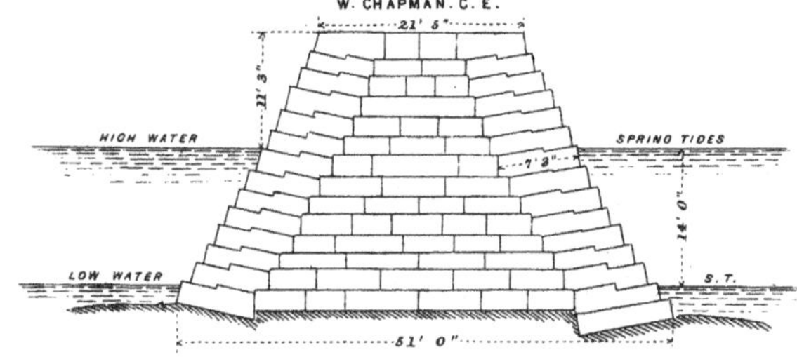

NORTH EAST PIER, SEAHAM.
W. CHAPMAN. C. E.

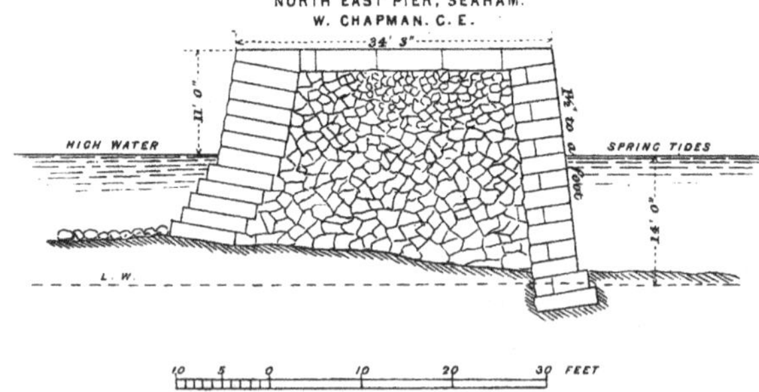

J. Bartholomew, Edin^r

Published by Adam & Charles Black, 1874.

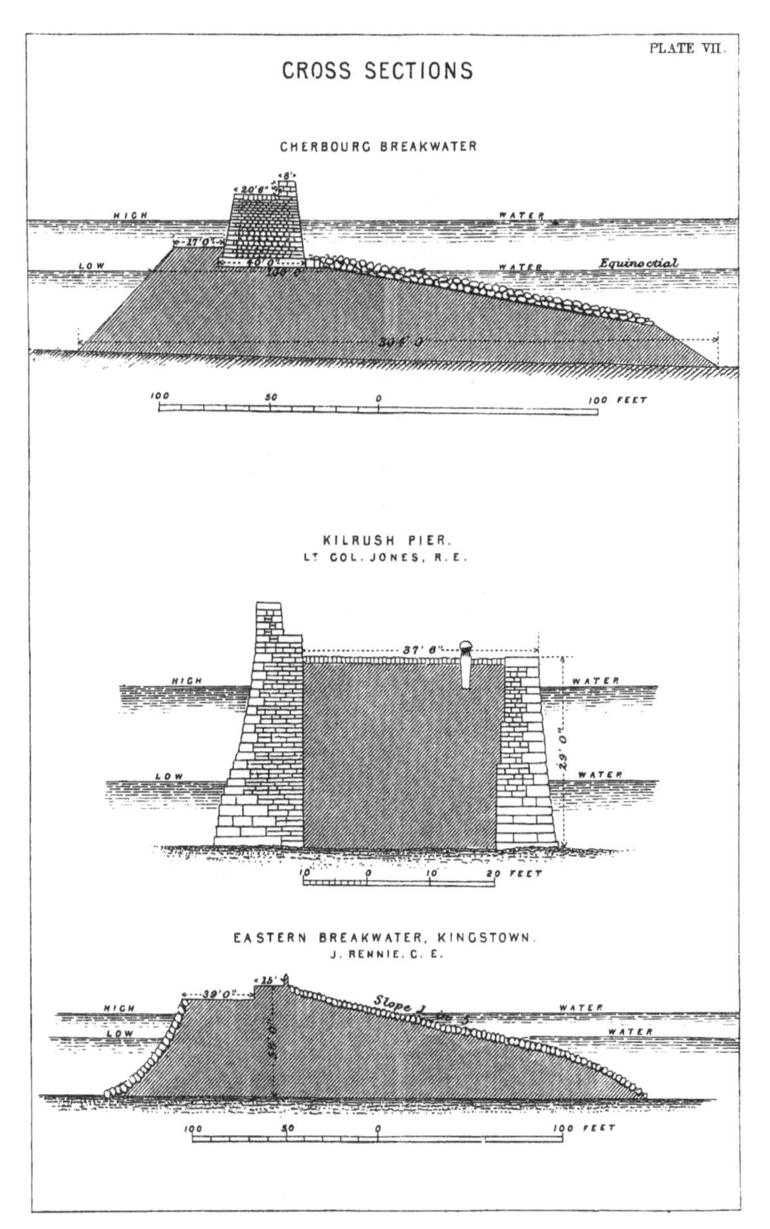

Published by Adam & Charles Black, 1874

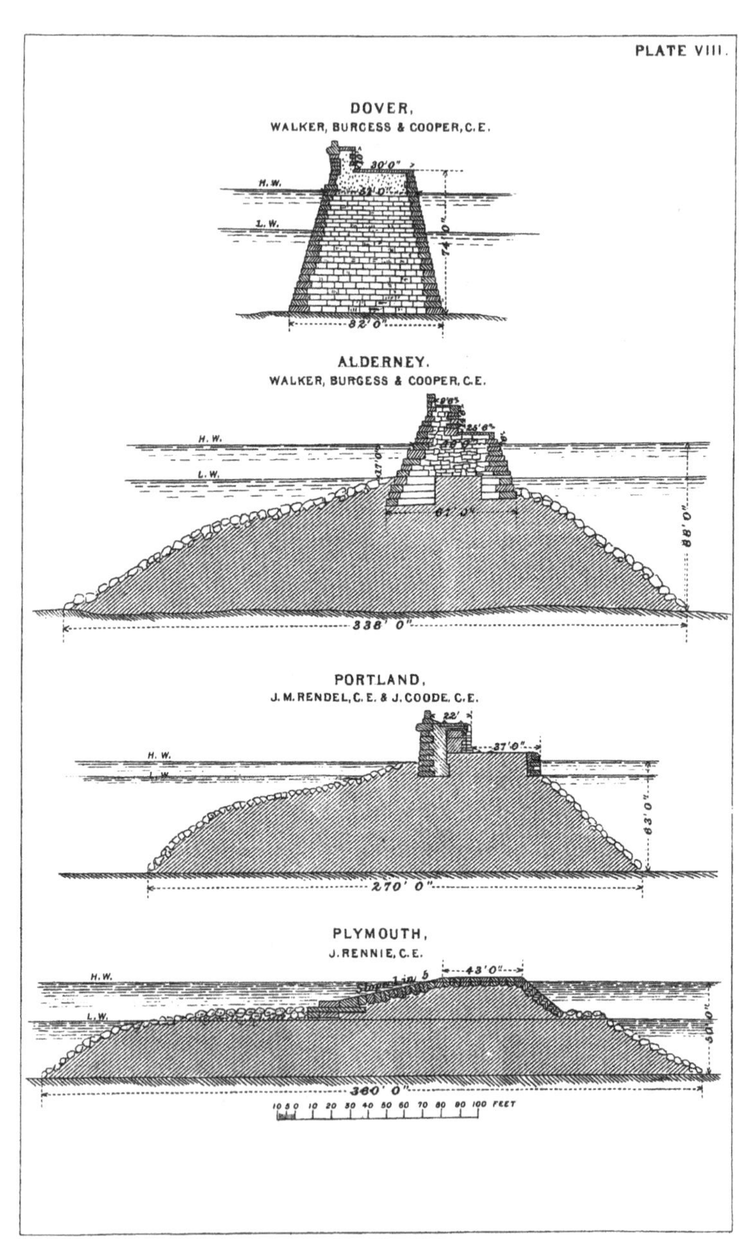

Published by Adam & Charles Black 1874.

PLATE IX.

JETTY OF ANCIENT PORT OF DUNKIRK.
1699.

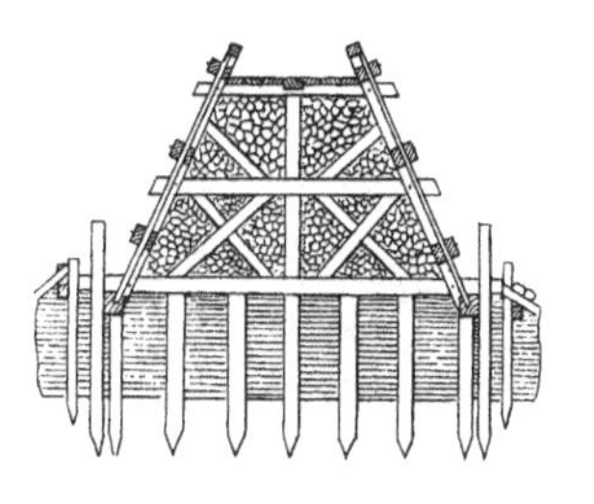

INVERGORDON JETTY.
D.& T. STEVENSON. C.E.
1856.

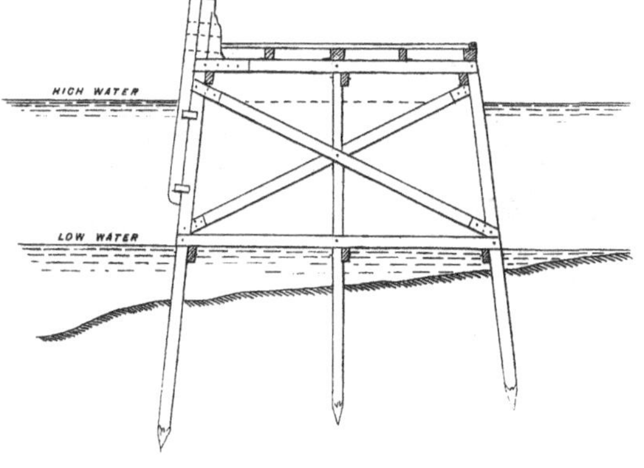

BLYTH BREAKWATER, J. ABERNETHY. C.E.
1856.

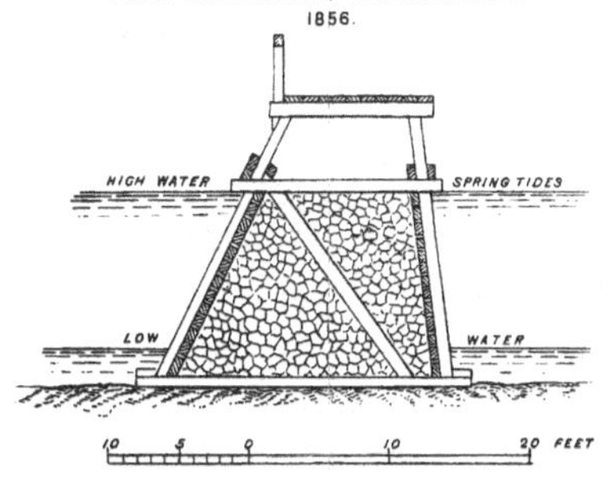

Published by Adam & Charles Black, 1874.

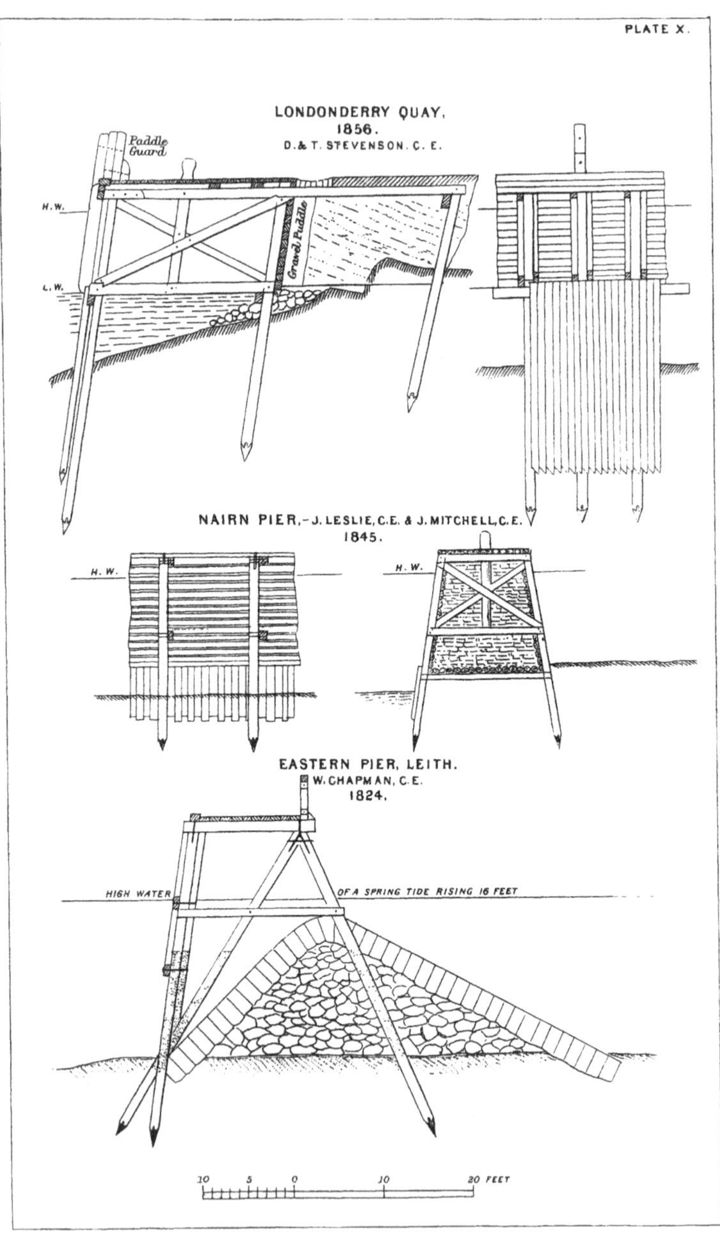

Published by Adam & Charles Black, 1874.

WICK BAY

PLATE XI.

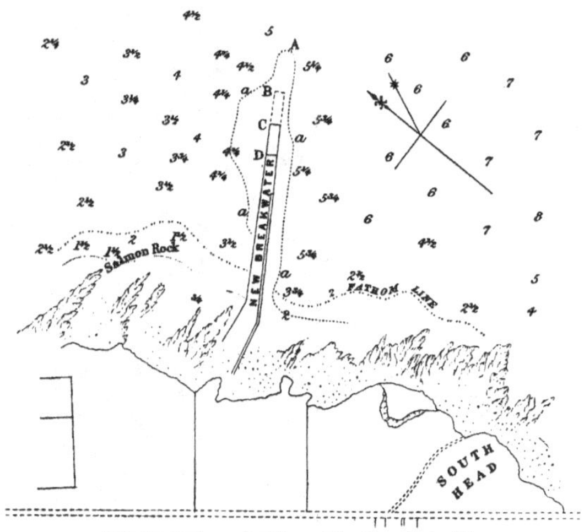

NOTE *The depths are in fathoms at Low Water Spring Tides.*

REFERENCES

A. *Extreme distance to which the Rubble base was carried.*
B. *Extreme distance to which the superstructure was carried.*
C. *Point to which the building was restored & finished by Cement Blocks.*
D. *Present end of building left standing.*
The dotted line a shews the extent of the Bay occupied by Rubble laid down from the Staging or spread by Storms as last ascertained after the damage of Feby 1870.

SCALE

100 50 0 100 200 300 400 500 600 700 800 900 1000 Feet

Elevation of the end of the Breakwater shewing in diagonal shade lines the Mass of 1350 tons which was removed entire and the manner in which it was connected together by iron bars.

SCALE

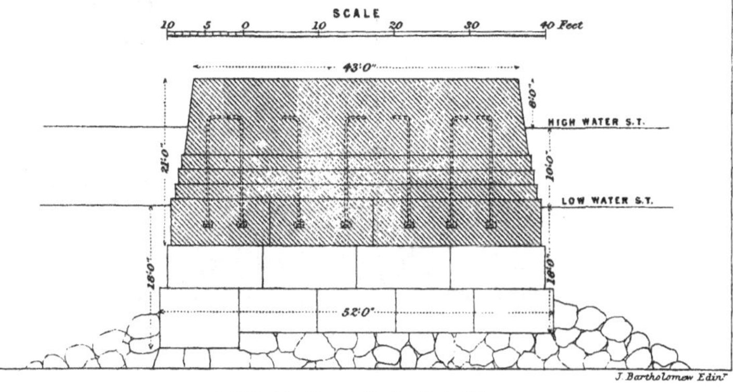

J. Bartholomew Edinr.

Published by Adam & Charles Black. 1874.

PLATE XII.

PLATE XIII

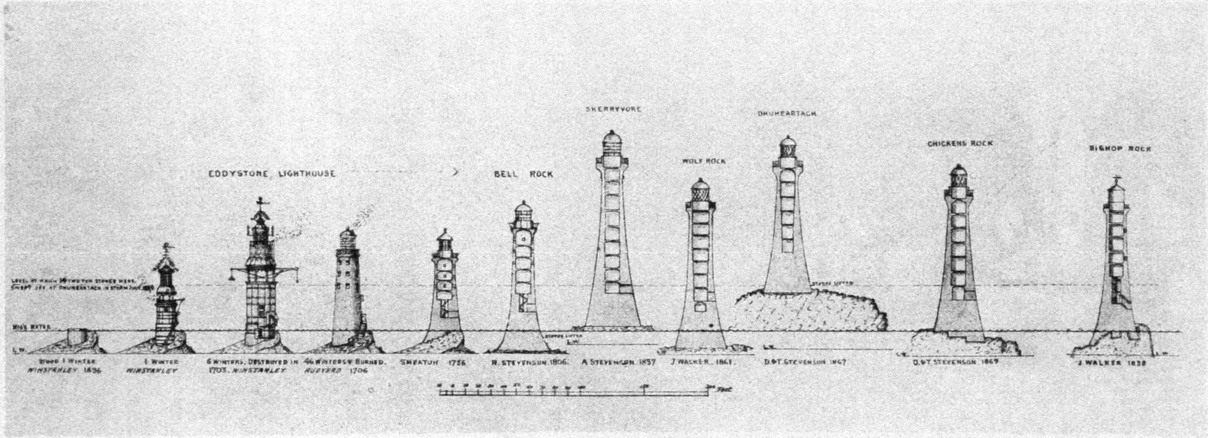

EDDYSTONE LIGHTHOUSE
BELL ROCK
SKERRYVORE
WOLF ROCK
DHUHEARTACH
CHICKENS ROCK
BISHOP ROCK
HIGH WATER
L.W.
WOOD 1 WINTER
WINSTANLEY 1696
1 WINTER
WINSTANLEY
6 WINTERS, DESTROYED IN
1703. WINSTANLEY
46 WINTERS & BURNED.
RUDYERD 1706
SMEATON 1756
R. STEVENSON 1806.
A STEVENSON 1857
J. WALKER. 1861.
D & T STEVENSON 1867
D & T STEVENSON 1869
J. WALKER 1858

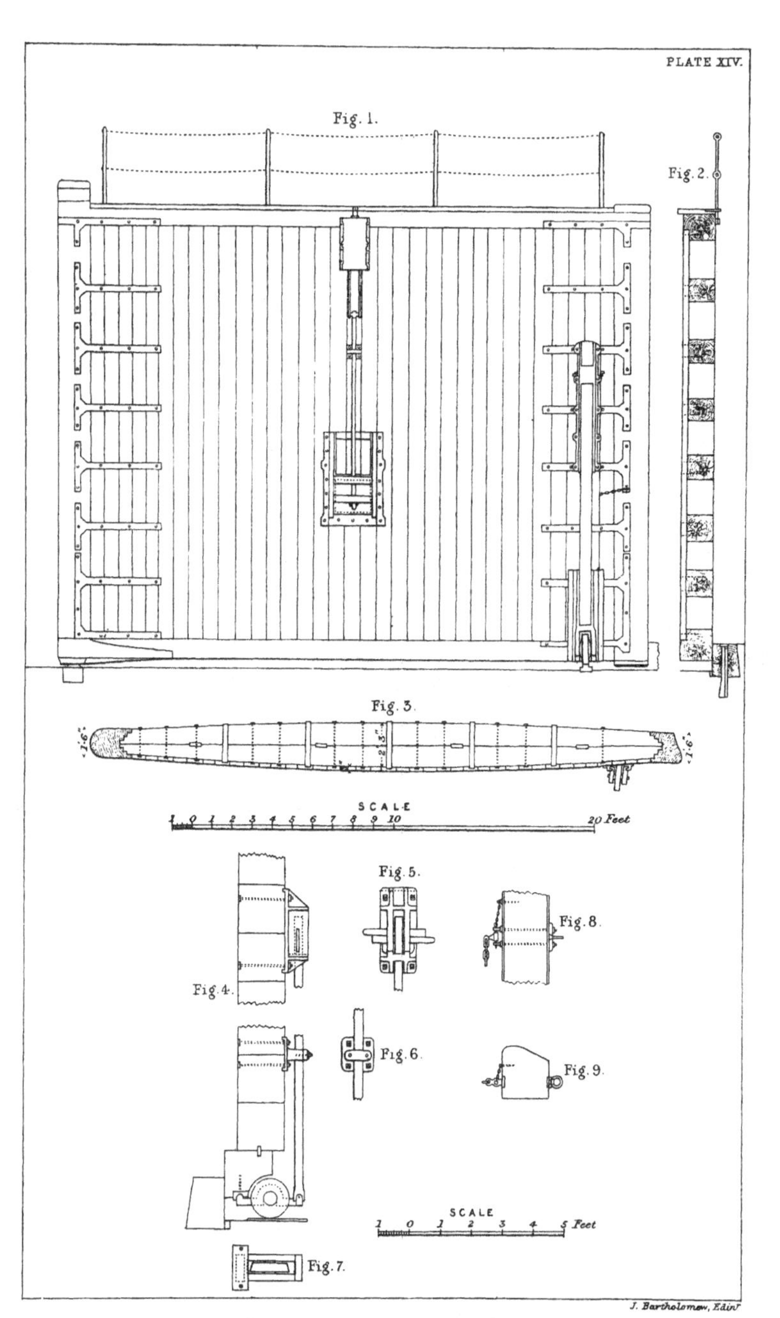

Published by Adam & Charles Black, 1874.

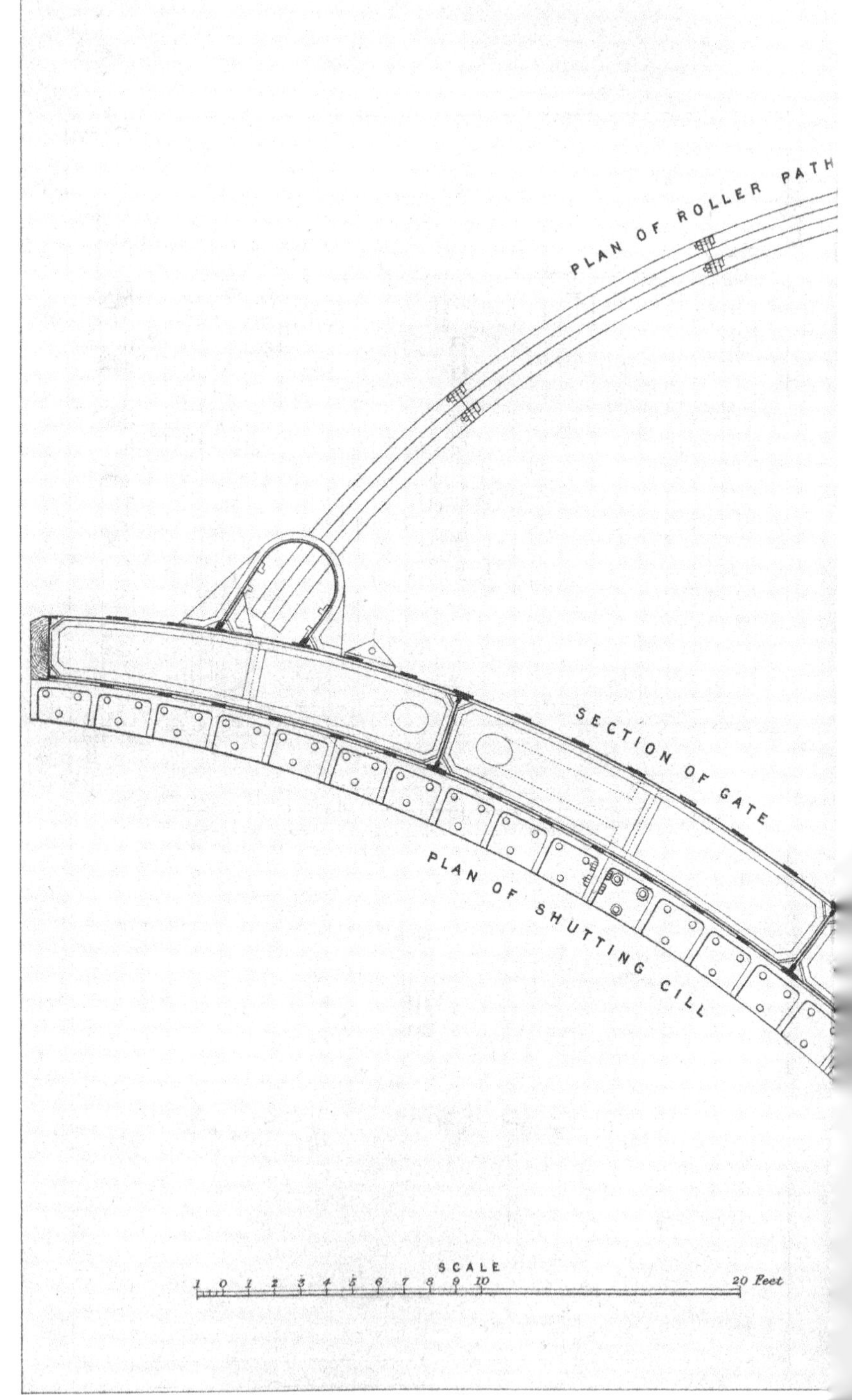

Published by Adam

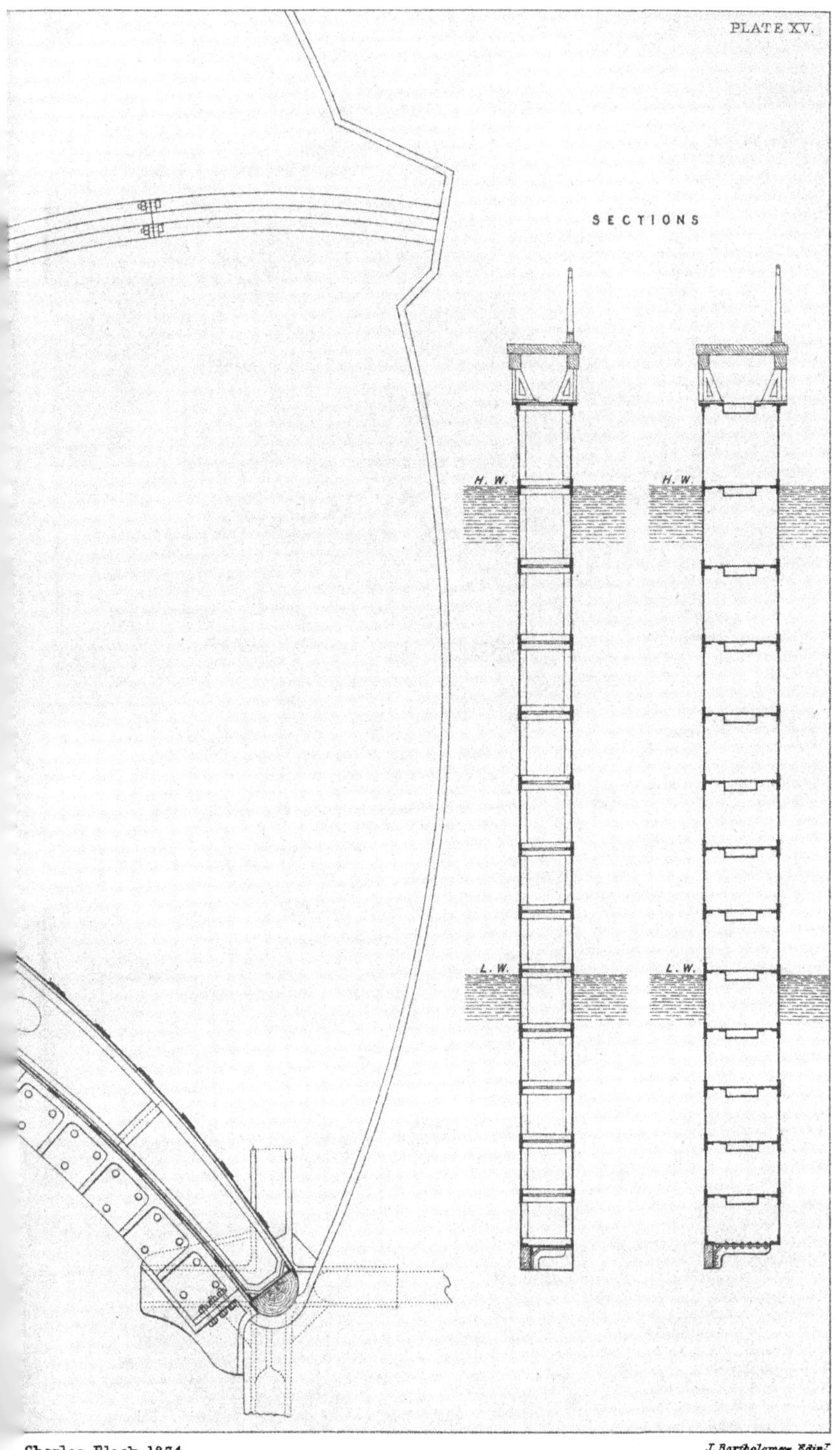

Charles Black, 1874.
J. Bartholomew, Edin.r

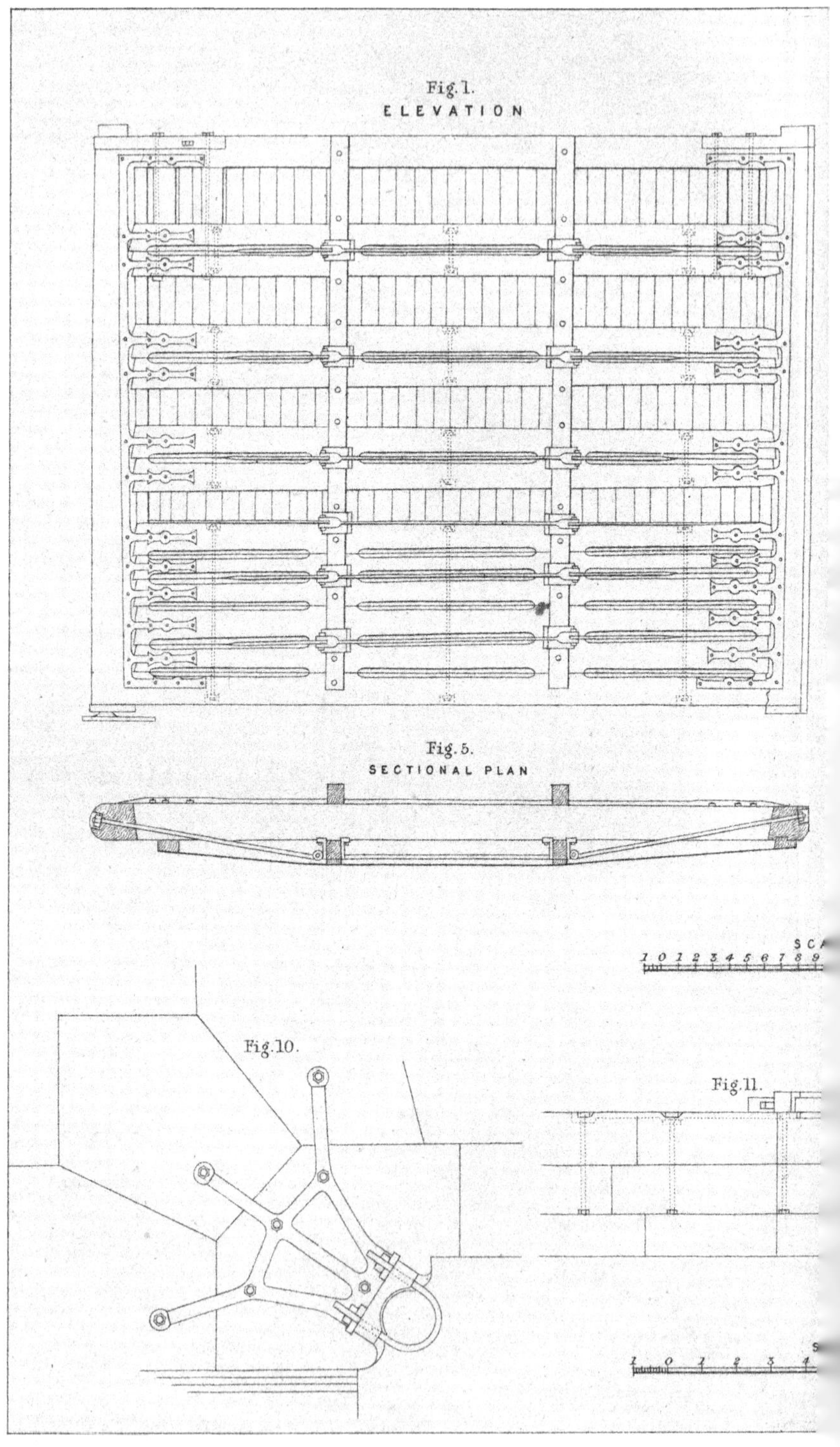

Published by Adam

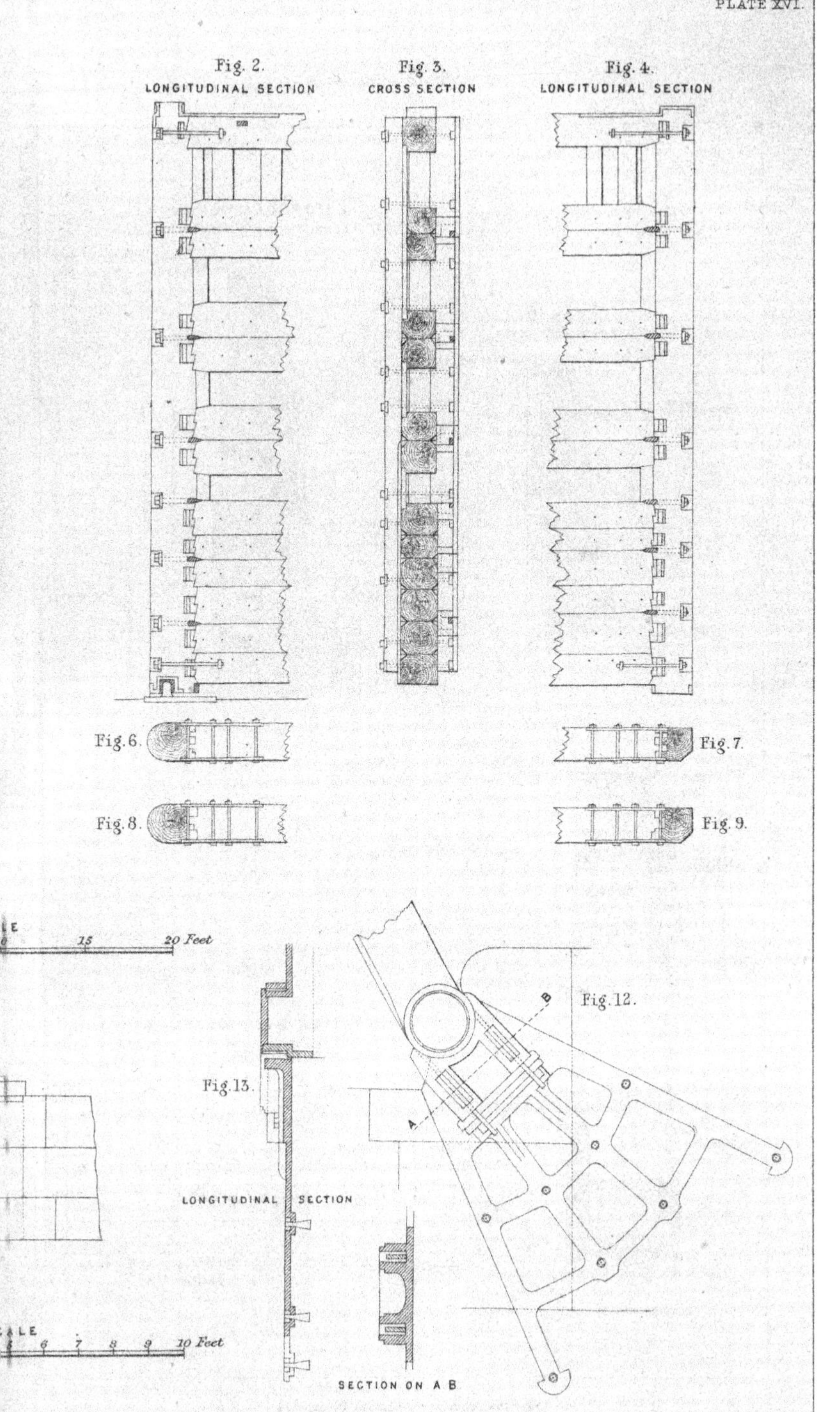

Charles Black, 1874.
J. Bartholomew, Edin.r

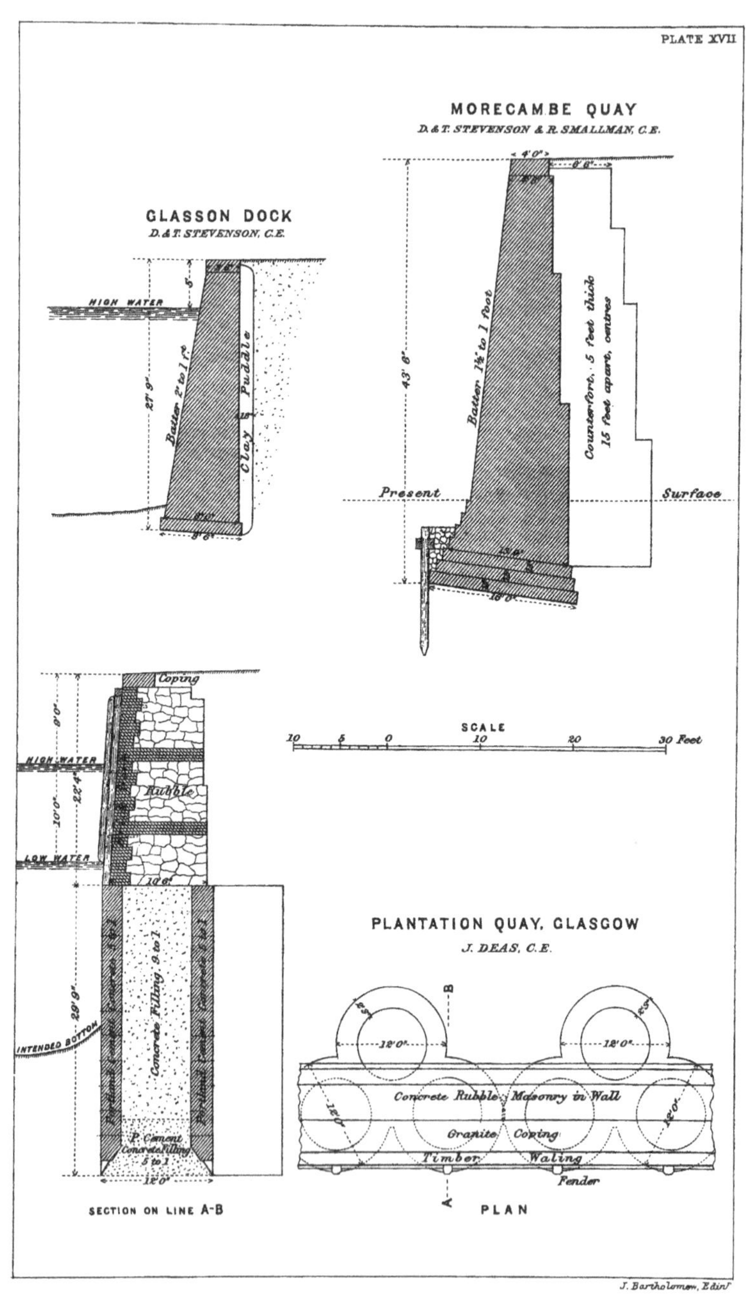

Published by Adam & Charles Black, 1874.

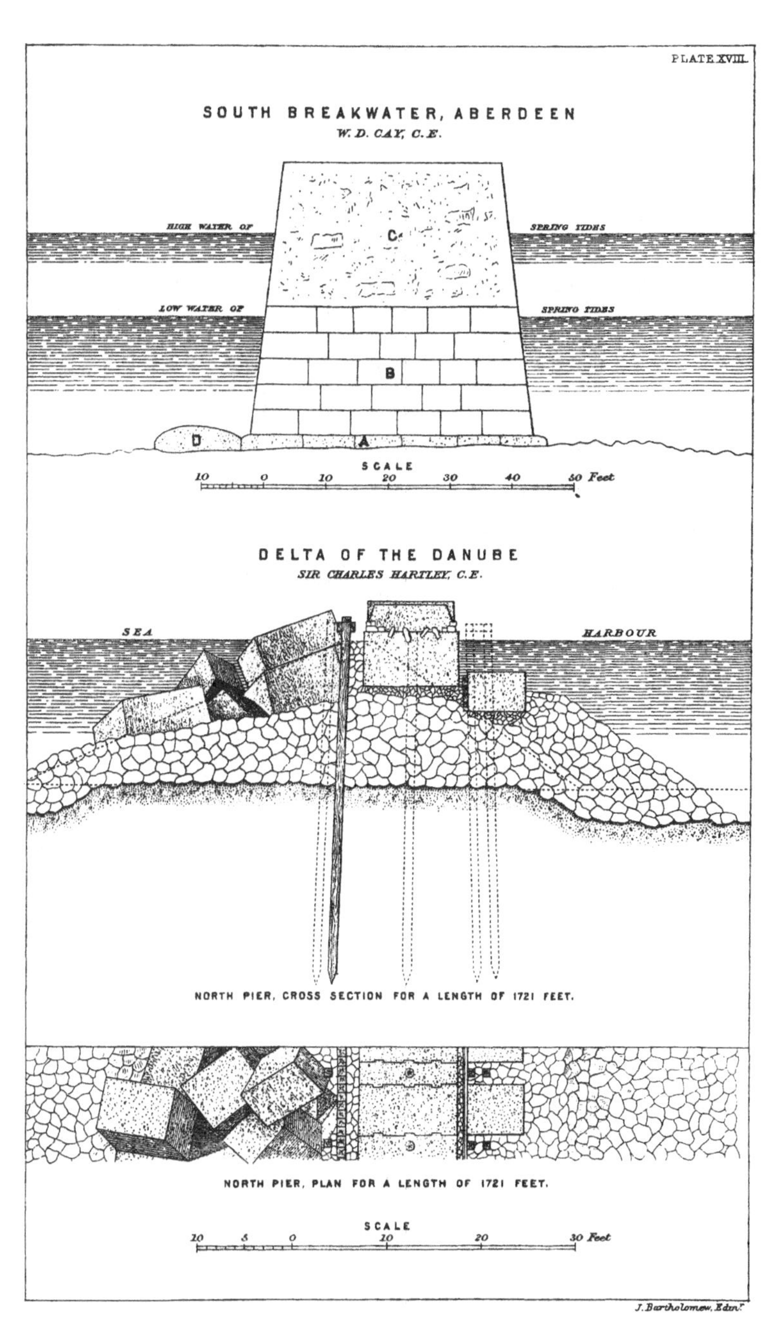

Published by Adam & Charles Black, 1874.

PLATE XIX.

ANSTRUTHER BREAKWATER

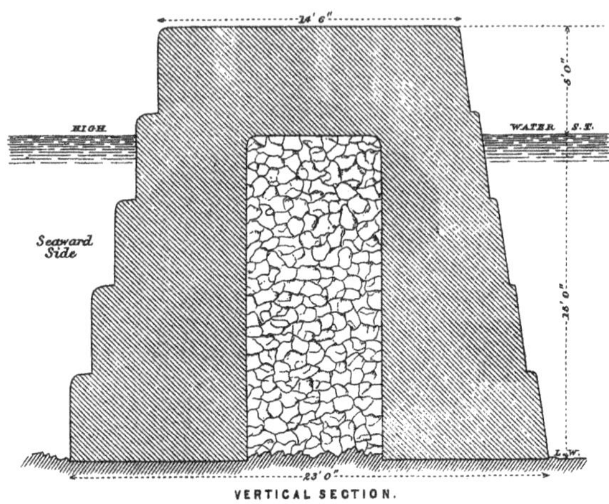

VERTICAL SECTION.

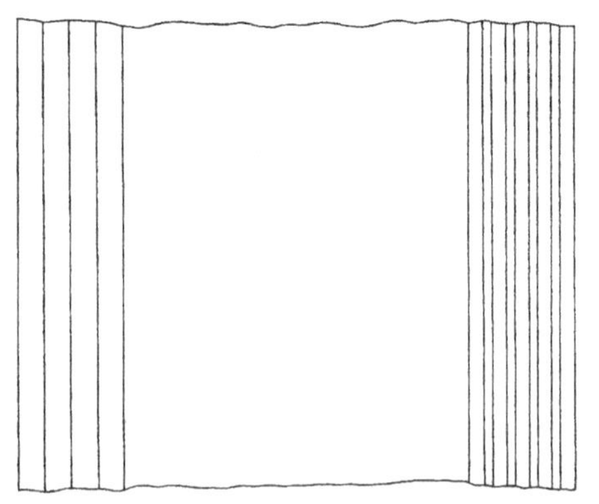

PLAN.

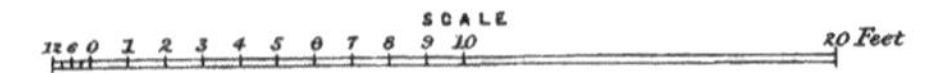

J. Bartholomew, Edinr.

Published by Adam & Charles Black, 1874.

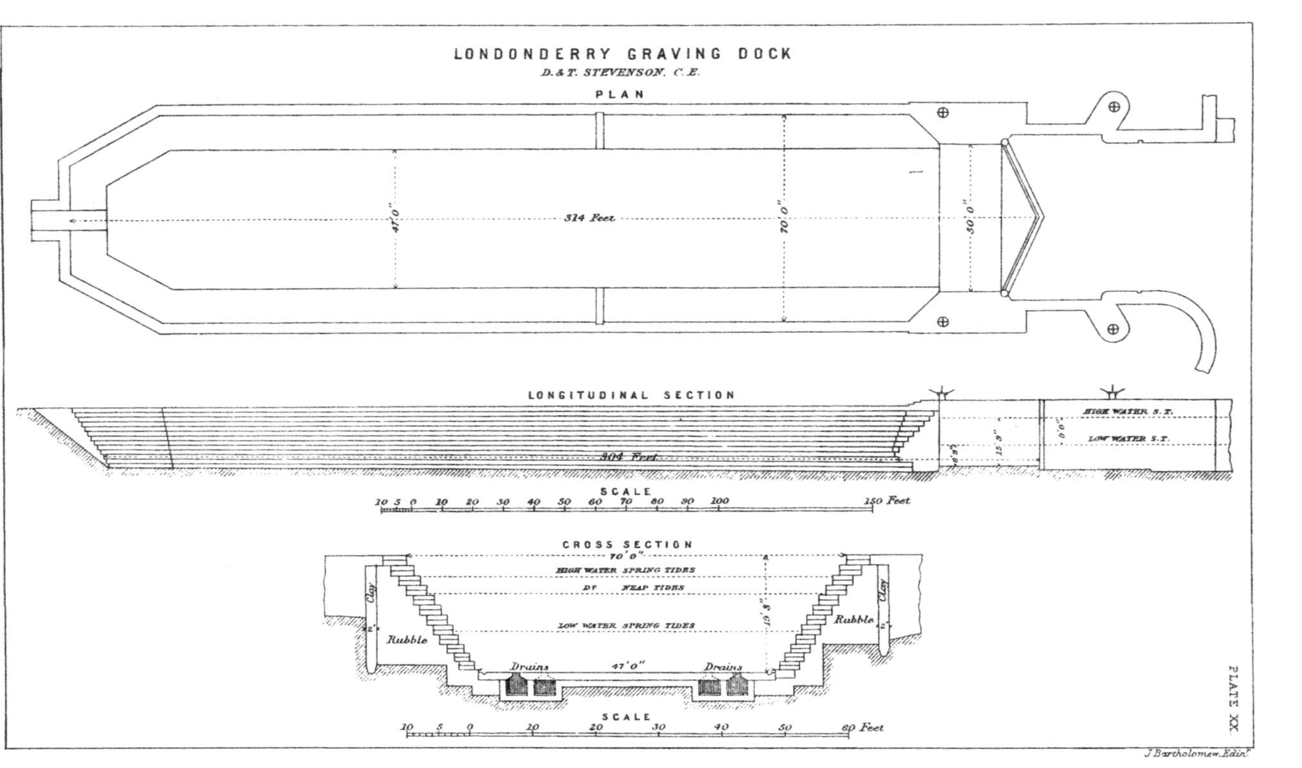

Published by Adam & Charles Black, 1847.

Printed in the United States
By Bookmasters